ECONOMICS OF MACHINE TOOL PROCUREMENT

By Wilbert Steffy
     Professor Emeritus
     Department of Industrial and
       Operations Engineering
     Research Engineer
     Industrial Development Division
     The University of Michigan
     Ann Arbor, Michigan

Copyright by
The Society of Manufacturing Engineers
All rights reserved.
Library of Congress Catalog Number 77-90986
International Standard Book Number 0-87263-041-2

The Society of Manufacturing Engineers
20501 Ford Road, P.O. Box 930, Dearborn, Michigan 48128

PREFACE

Although fundamental management has changed little in the past decade, there have been numerous changes in the scope and complexity of management decisions. Important management changes in recent years have been the trends toward automaticity, the development of computer technology and the demand for higher technical and business training in management personnel.

This book is written primarily for those individuals who have to cope with the enormous problems of capital spending where complex, automatic and costly equipment is involved.

This book, therefore, is directed towards production managers, design engineers, sales engineers, financial managers and educators involved in teaching Engineering Economics, Financial Control, Methods Engineering, and Cost Accounting. Specially, it is designed to help the individuals who are responsible for capital decisions in the area of machine procurement.

The engineer, for example, enjoys his ability to manipulate materials, processes and machines. But he may be poorly trained in costs and engineering economics. This weakness is being remedied in many educational institutions. This is necessary because many companies are demanding that their engineers be thoroughly familiar with the costs entering production, including machining costs. In addition, they are asked to be knowledgeable in the various financial transactions, including the source of money and the various payments, necessary in a business enterprise. These trends are making the engineer-manager aware of higher-level decision-making processes.

The present book is designed to help the engineer-manager be comfortable in the areas indicated above. With this in mind, the book has been divided into the following areas:
1. Numerical Control Generalities
2. Simple and Compound Interest
3. Depreciation Methods and Profit Based Taxes
4. Capital Budgeting Decisions
5. Rate of Return Methods
6. Cash Flow Management
7. Various Replacement Algorithms
8. Additional Decision Making Tools

Numerical Control Generalities discusses the various pro's and con's of automaticity. It attempts to provide sound advice to prospective investors in this type of equipment.

The *Simple and Compound Interest* section is designed to acquaint the reader with the various interest methods in use for making sound capital equipment investments. The procedures of discounting and compounding are discussed in some detail.

Depreciation Methods are discussed from the viewpoint of the decision faced by management in determining which method to use. It has been found that methods which provide better market value estimates involve a greater accounting cost. The effect of the depreciation method choice on profit based taxes is featured in this section.

The *Capital Budgeting* chapter covers the capital projects that must be presented to management. The funds generated internally and the loans that may be necessary are discussed with detailed examples.

The *Rate of Return* chapter presents various rate of return methods that are in use today. However some are more popular than others. All the available methods are discussed in detail with appropriate quantitative examples.

The several *Replacement Algorithms* in this chapter were developed from actual examples. The detailed data were collected from various corporations located in the Mid-West area. The Industrial Development Division of The University of Michigan used several of these examples in some of their own publications.

Additional Decision Making Tools are presented to stimulate the reader to do further research in this very important area of decision making. It is hoped that this book will provide that stimulation.

Wilbert Steffy
Ann Arbor, Michigan

CONTENTS

CHAPTER 1 INTRODUCTION 1

CHAPTER 2 SIMPLE AND COMPOUND INTEREST 7

CHAPTER 3 THE DEPRECIATION METHOD DECISION 29

CHAPTER 4 DECISIONS OF CAPITAL BUDGETING 53

CHAPTER 5 RATE OF RETURN PRINCIPLES 63

CHAPTER 6 CASH FLOW MANAGEMENT 71

CHAPTER 7 THE GENERAL REPLACEMENT ALGORITHM 83

CHAPTER 8 OTHER INVESTMENT CRITERIA 107

Society of Manufacturing Engineers
20501 Ford Road, P.O. Box 930, Dearborn, Michigan 48128

INTRODUCTION

CHAPTER 1

Numerical control has been somewhat extravagantly described as the Second Industrial Revolution. While this terminology may be justified, the individual company considering a numerical control investment must be cautious. Numerical control is not best for everyone. Furthermore, those who are likely candidates for numerical control still face important and difficult decisions as to which NC unit to buy. The large capital investment required for a numerical control operation is ample reason to conduct a careful cost analysis before ordering equipment. However, most companies have neither the time nor the data required for an intensive analysis.

The cost effects of numerical control which deserve attention in careful analysis are numerous and frequently ambiguous. Simple answers do not exist. This book will attempt to provide sound advice for prospective investors in numerically controlled metalworking equipment.

Procedure

A question that must be answered is whether there is any difference between the procedure for evaluating numerical control and the procedure for evaluating any other capital expenditure. A qualified "no" would seem to be the answer. The analytical considerations applicable to any capital investment are also applicable to numerical control. Capital recovery costs, operating costs, maintenance costs, and productivity comparisons are relevant aspects of any machinery procurement decision.

Experts on investment analysis repeatedly state that only those matters that are different in the prospective investment alternatives, are relevant to their comparison. In evaluating conventional machinery, many cost elements can be regarded as common to all alternatives. Such indirect costs as inventory, material handling, tooling, and inspection vary little, regardless of which machine is selected.

When numerical control is considered as an alternative investment, however, the situation changes. In fact, very few manufacturing costs remain unchanged. The problem is compounded by the limited amount of data available for measuring these changes. To make an intelligent investment decision, the investment analyst needs more than a tabular routine. He also needs aids or guidelines to estimate the effect of numerical control on manufacturing costs.

In any capital equipment evaluation analysis, two series of costs must be considered. The first is known as the capital recovery costs, *i.e.*, the recovery of the initial investment for the new equipment. The second series is the annual operating and maintenance costs. Typically, these costs increase each year as the maintenance costs increase with the age of the machine. The rate at which capital costs are recovered depends upon the method of depreciation used.

The graph in *Figure* 1-1 shows the two series of costs plotted over time. The goal of the analyst is to determine whether the net savings in operating and maintenance costs with numerical control are enough to more than offset the higher initial outlay. If they are, the minimum in a cost graph for a numerically controlled machine will be lower than the minimum for the conventional machine with which it is being compared.

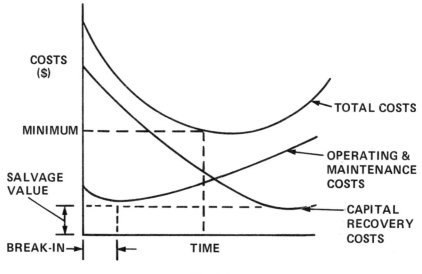

Fig. 1-1

The initial investment includes all costs which are necessary to get a numerically controlled machine working. Among these costs are the purchase price of the machine; shipping and installation costs; the cost of training operators, maintenance and programming personnel; and the costs of any necessary tools and equipment.

The net savings between the two methods can be divided into two categories, direct savings and indirect savings. Direct savings refers to the savings in machining time by using numerical control, i.e., the hours it takes to complete jobs on a conventional machine minus the time it would take to machine the same jobs on numerical control. This savings can be converted to a dollar figure by using an appropriate rate such as the machine operator's pay rate.

Indirect savings occur in many facets of the total machining costs. These include: maintenance, tooling, programming (always a negative savings because there is no programming on conventional machines), inspection, scrap, inventory, floor space, and material handling.

Discounted Cash Flow

The various costs and savings are evaluated by means of the technique of discounted cash flow, which recognizes the time value of money. A dollar received today is more valuable than a dollar received a year from now. This is because a dollar today can be invested, and there is some risk that the dollar a year from now will not be received. The cash flows (initial investment and

INTRODUCTION

annual costs and savings) are discounted, and an interest rate representing the rate of return is computed. This rate is then compared with the minimum rate which the firm considers acceptable for a capital investment. The minimum, or hurdle, rate includes the cost of capital funds and the firm's willingness to assume risk.

The method of depreciation affects the after-tax cash inflows because depreciation charges are tax-deductible expenses. Depreciation is the process of allocating the initial costs of a project over the useful life of the project. The higher the depreciation charge that can be allocated to any given year, the lower the taxable income for that year.

A simplified example will help to illustrate how the discounted cash flow technique is used to compute the rate of return. Suppose a firm is considering a numerically controlled machine which costs $30,000 and has a ten-year estimated life with zero salvage. Training cost for personnel is $2000, shipping and installation is $1000, and additional equipment is $7000. The direct savings, determined by an analysis using the methodology, is 950 hours per year. If the operator is paid $4.00 per hour, the direct savings amounts to $3800 per year. The indirect savings, again found by using the methodology, are as follows:

Maintenance	−$ 400
Tooling	2000
Programming	− 2500
Inspection	400
Scrap	500
Inventory	1200
Net Indirect Savings:	$1200 per year

The total annual savings is then $5000 per year. If the tax rate is 50 percent, the after-tax savings is $2500 per year.

Straight-line Depreciation

In order to account for the tax deductible feature of capital investment through depreciation charges, one must select a recognized depreciation method. One common method of depreciation is called straight–line depreciation. Each year an equal portion of the initial investment is subtracted as expense for the year. The net effect is an annual cash inflow equal to the depreciation charge times the tax rate. In this example, assuming a ten-year machine life and a $30,000 machine cost with zero salvage the after-tax cash inflow is $3000 × 0.50 or $1500. Adding this to the after-tax savings ($2500) yields a net annual cash inflow of $4000. The discounted cash flow technique will determine the interest rate at which $4000 per year for 10 years will be equivalent to $30,000 today. The diagram below shows these cash flows.

ECONOMICS OF MACHINE TOOL PROCUREMENT

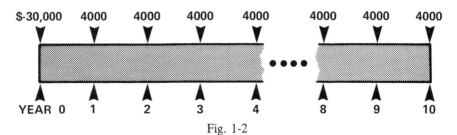

Fig. 1-2

Using compound interest tables, the analyst determines what interest rate will make a series of $4000 investments for 10 years equal to an initial sum of $30,000. At that interest rate the "present value" of savings series is $30,000. The rate for this problem is slightly more than 5½ percent.

Data collection remains a difficult problem for any quantitative approach. This problem is particularly evident in firms considering their first NC purchase. One serious study, in effect, recommends *all* metalworking firms purchase at least one NC Unit just for the "experience," charging any mistakes off to research and development. Considering a typical installation often costs upwards of $200,000, a trial and error approach seems hardly appropriate. Nevertheless (although few firms will admit it), this approach has become very popular.

A more refined method of cost analysis suggests collecting detailed cost data on a small sample of typical workpieces. One sample set is run on conventional machines; a second set is run on the NC unit under study; the resulting data are then compared. Procedural guides for such a cost analysis are being distributed to industry on pretty much a "do-it-yourself" basis. Several assumptions behind this approach are questionable, however.
1. It assumes that all firms have access to an operating NC installation of their choice during the NC test phase.
2. It assumes that the operating data obtained are not only typical under existing conditions, but also constant over the life of the machine.
3. It assumes no relationship between risk and time.
4. It assumes industry will initiate an effort to methodically collect required data and perform necessary calculations.

A description of several parameters will illustrate this point. There is general agreement that NC machines, particularly NC drilling machines, have proven themselves moderately, if not sensationally, economical on small batches of highly complex parts. Another general opinion is that high operator wage rates increase the attractiveness of numerical control. Thus three important parameters appear to be batch size, workpiece complexity, and operator wage rate. In analysis of a particular NC investment decision, emphasis should be on determining the relative effects on total cost as important parameters are varied within their expected range.

NUMERICAL CONTROL LITERATURE

A discussion of the current status of numerical control literature serves as a worthwhile preview to the methodology developed in this study. Although the author recognizes the tendency of researchers to underrate the work of others,

INTRODUCTION

he still maintains that current literature on NC justification fails to provide the potential numerical control buyer with adequate information.

Most of the literature on numerical control justification is qualitative. Advantages of numerical control are typically listed along with lists of manufacturing conditions necessary to achieve these advantages. Familiarity with this qualitative literature is a must to anyone considering numerical control. However, before a procurement decision is made, the attributes of numerical control must be measured.

Some attempts have been made to supply numerical control buyers with cost figures to justify their decision to invest in numerical control. One form of information permeating numerical control literature consists of statements such as "Company X saved $Y\%$ of their direct costs with a brand Z numerically controlled machine." In reading such statements, one finds it difficult to avoid recalling the familiar toothpaste commercials (... $Y\%$ few cavities with ...).

While this approach may be sufficient inducement to purchase toothpaste, it is hoped that a numerical control procurement decision can be based on more relevant criteria. Numerical control success stories, even those that are well documented, provide a risky basis for numerical control justification. What is good for company X may be disastrous for other companies. The wide variance of success (or failure) of numerical control is attested to by one author's statement:

"The claims made for numerically controlled manufacturing range all the way from 0 to 99 percent savings in costs. In contrast, some manufacturers have actually experienced increased costs in utilizing numerical control."

Because of the wide variance in costs, each numerical control procurement decision should be based on data and circumstances relevant to it alone.

The task of performing a numerical control investment analysis has been discussed in several publications. Procedures for evaluating numerical control have been outlined. Typically, the procedures have been demonstrated on real or hypothetical case data, and forms have often been provided for the reader's own analysis.

The payback period method of evaluation is used in most of the current publications. The author believes, however, that an alternative method of evaluation—the rate of return, for example—is more favorable. A second criticism is more important and represents the major reason for writing this book. Most authors of past publications have failed to adequately examine the nature of the data available to numerical control investment analysts. Indeed, the questions of data availability is often overlooked in order to present a neat, concise methodology. While this approach may provide an analysis procedure that looks favorable initially, the practitioner is likely to find it difficult, and often impossible, to supply inputs to the analysis that are anything better than very rough approximations. It is recognized that an investment analysis cannot avoid a certain amount of guesswork, but the author believes that improvements can be made in both the nature of required data, and in the amount of information placed at the analyst's immediate disposal.

Before various replacement algorithms are discussed in detail with appropriate examples and using the recommended rate of return method, it is appropriate that some of the details involved in presenting the cost models be discussed. Thus, simple and compound interest will be presented in some detail, equivalence will be defined with examples, depreciation methods will be explained with examples and graphs, and capital budgeting, tax effects, and cash flow will be integrated into the overall decision problem. Finally, a detailed presentation of the "Rate of Return" method will be given.

CHAPTER 2

SIMPLE AND COMPOUND INTEREST

The classic economists have traditionally defined interest as a reward for the use of capital in the production of goods and services. Similarly, wages were defined as labor's reward for its contribution to the production process. And rent was defined as the reward for land's contribution. The percentages of money received for these three contributing elements to the production process varied considerably in a competitive market, but they generally followed the law of supply and demand.

Today, however, these concepts may no longer be valid. Wages in a great many companies are determined by negotiated contracts between labor unions and company management. Rents, especially during periods of emergency, are controlled by regulatory government agencies. Likewise, capital rewards are often limited by laws and regulations, especially in the public utility field.

Money can be considered valuable only in the sense that it can be used as a medium of exchange for goods and services. As such, it represents a certain number of units of goods and/or services. Thus, money can be considered a warehouse or reservoir from which withdrawals can be made for the future purchase of goods and services. In this respect, money itself renders a unique service as a medium for exchange and, therefore, a price must be paid for this service. In this respect, it commands a price which is dependent on supply and demand. The price paid for its use is governed by its value at that point in time—and upon the risks assumed by the lender of money for foregoing the services that his money can buy over some future time period.

For example, the lender of money may demand a higher than normal price when he feels that the probability of repayment is low. Conversely, the price for the use of money decreases when the borrower puts up sufficient collateral. Thus risk and the law of supply and demand determine how much a borrower must pay for the use of someone else's money.

This payment for the use of other people's money is called interest. Some writers define interest as a consideration for the use of capital. Others define interest as the time value of money. In any event, it is something that one must pay if borrowing is necessary. Interest could easily be considered as rent paid for the use of money in much the same way that rent is paid for the use of a house or lot.

Simple Interest

When money is used as the common denominator for various types of transactions, several mathematical formulas can be used to show relationships between the amount of money earned by the lender to the total amount of money lent to a borrower. Problems involving the use of money for simple interest purposes have five fundamental elements, as follows:

1. Principal—Basic amount of money on which interest is paid.
2. Interest—The price paid for the use of the principal amount.

3. Time—The days, weeks, months, or years of use for the principal amount. It is generally expressed as a percentage rate per year.
4. Rate—Generally expressed as a percent of the principal, rate is the price paid for a unit of time.
5. Amount—The amount is the principal plus interest for the time period stated.

Where simple interest problems are involved, the interest earned is directly proportional to time. In order to express money relationships mathematically, let:

P = Principal or present worth in dollars.
I = Total interest in dollars.
r = Simple interest rate expressed as a percent.
t = Time in years or fractions of years.
s = Final amount in dollars.

Expressing simple interest as a formula:

$$I = Prt \quad\quad (1)$$

Graphically this simple interest relationship can be shown where the "final amount" is the ordinate and "time" is the abscissa.

GRAPHICAL REPRESENTATION OF SIMPLE INTEREST

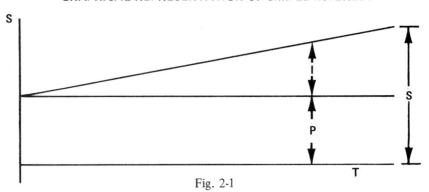

Fig. 2-1

The interest rate r is expressed as a percent rate per unit of principal per unit of time.

The final amount s is:

$$s = P + I$$

where:

$$I = Prt$$

Then,

$$s = P + Prt$$
$$= P(1 + rt) \quad\quad (2)$$

Formula 2 states that the principal P will grow to s in t years with a simple interest rate of $r\%$ per year.

SIMPLE AND COMPOUND INTEREST

Conversely, the present worth of a sum s over time t at r rate is equal to:

$$P = \frac{s}{(1 + rt)} \quad \quad (3)$$

These simple interest equations can be illustrated as follows: Assume that a borrower receives a simple interest loan of $100 for three months at 5 percent simple interest.
1. How much interest is earned?
2. What is the total amount to be repaid?

Use Equation 1:

$$I = Prt$$
$$= (100)\ (0.05)\ (3/12) = \$1.25 \text{ interest earned.}$$

Use Equation 2:

$$s = P + I$$
$$= 100 + 1.25 = \$101.25$$

Or

$$s = P + Prt$$
$$= 100 + (100)\ (0.05)\ (3/12) = \$101.25$$

Simple Discount

In discussing simple interest it has been noted that the interest is considered as a fraction of the principal. But with simple discount we consider the interest as a fraction of the final amount. This is the basic difference between simple interest and simple discount. Essentially, simple discount is the process of finding the present worth of a future sum and can simply be called discounting the future sum. Thus, the amount of money representing the difference between the future sum and the present worth of this future sum is the interest representing the discount on this future sum. The ratio of this difference expressed as a percent with the future sum is called the discount rate. An example will make this relationship clear.

Let:
- D = Total discount in dollars.
- s = Future amount to be discounted in dollars.
- d = Rate of discount expressed as percent.
- n = Number of years.
- P = Present amount in dollars.

Then:

$$D = sdn \quad \quad (4)$$

Also:

$$P = s - sdn$$
$$= s(1 - dn) \quad \quad (5)$$

Suppose a borrower contracts to pay back $100 in three months at 4 percent simple discount.
1. How much will the future be discounted?
2. How much will the borrower receive?

Using Equation 4:
$$D = sdn$$
$$= 100\,(0.04)\,(3/12) = \$1.00$$

Using Equation 5:
$$P = s - sdn$$
$$= 100 - [(100) \times (0.04) \times (3/12)] = \$99.00$$

This amount is the present worth of $100 with a 4 percent simple discount rate.

Lending institutions sometimes employ simple discount in lending small sums of money for short periods of time. When this is done, the interest is paid in advance. For example, a borrower may wish to borrow $100 from a bank. Using simple discounts, the bank states that the borrower must pay to the bank $100 at the end of the year. It then gives the lender a sum equal to the present worth of the future sum of $100 one year hence. If the bank indicates that the simple discount rate is 4 percent, the borrower receives $96, *i.e.*, 100 − (0.04)(100). Essentially, this is paying $4 interest to the bank in advance. Looking at this transaction from the present worth point of view, the borrower is actually paying slightly more than 4 percent. The calculation is:

$$\frac{4.00}{96.00} \times 100 = 4.17\%$$

The purchase or sale of goods is another area where simple discount is frequently used. The sales contract determines the particular conditions under which both parties will work. A great many contracts for the sale of goods state that upon delivery of goods, payment must be made within 30 days with 2 percent discount on the sales price if paid within 10 days. Let us examine a sale of $100, the agreed price, with stated payment terms of the rendered bill of 2 percent—ten days—net 30 days. By analyzing this transaction the following conditions can be presented for discussion:

Condition 1

(The buyer is B, the seller is S.)

B owes S $100 payable in 30 days. B has not only the use of the goods that S sold him, but he also has the use of $100 of S's money as represented in the goods sold to B with no interest except that which may have been applied to the price of the goods.

Condition 2

B can pay S the sum of $98 anytime within 10 days. B can, therefore, have the use of $100 for 10 days without paying any interest but will have to pay $2 for its use during the last 20 days of the 30-day period. Stated another way, B can receive $2 for the use of $98 for 20 days.

Condition 3

The $98 that S agrees to accept represents to him the present worth of $100 due 20 days later. The simple discount rate d is:

$$d = \frac{\$100 \ (0.02)}{\$100 \ (20/360)} = 36\%$$

If B takes advantage of the discount, he pays essentially $2 for the use of $98 for 20 days. The simple interest rate is:

$$r = \frac{\$100 \ (0.02)}{\$98 \ (20/360)} = 36.8\%$$

Note: The convention of 30 days per month is used for the above two calculations.

Compound Interest

We have seen that in the case of simple interest the principal remains fixed throughout the time cycle on which the interest is calculated. In the case of compound interest this satement is not true. Compound interest differs from simple interest in that the interest earned or payable at stated time intervals is added to the original principal. Then the interest for the next time interval is calculated on the sum of the principal and interest earned in the previous period. This act of adding the interest to the principal is termed compounding or converting the interest into principal. The time that elapses between two successive additions to principal is called the conversion or interest period. This period can be of any stated duration but is usually stated in terms of one month, three months, six months, or twelve months.

In *Figure 2-2* the letter P represents the principal or present worth and S represents the final amount after four conversion periods. The vertical line QK represents the compound amount of interest earned through four conversion periods. This line consists essentially of simple interest amounts calculated from bases other than the original base measured from the line AQ. These additional bases are represented by the lines $CO, EN,$ and GV. The simple interest line measured from base AQ is represented by line ACM, the simple interest line measured from base CO is represented by line CEL, etc.

The vertical line BC represents the simple interest on the fixed amount of principal P at the first conversion point. Compounding of interest begins at this point. The base for converting is changed from AQ to CO. The effect is to add the increment DE to the simple interest amount YD at conversion point 2. This process of converting interest into principal at each conversion point is continued until the total number of conversion periods have been covered.

Mathematically, the operation just described can be illustrated as follows:
Let

S_n = Future amount
P = Initial amount
i = Interest rate for each conversion period

OPERATION OF COMPOUND INTEREST

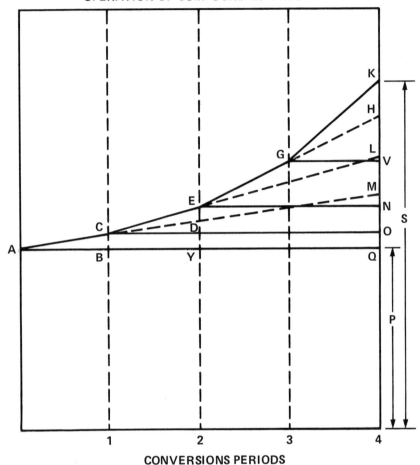

Fig. 2-2

Then the first

$S_1 = P + Pi = P(1+i)$ —(First conversion period)

Note: $Pi = BC$ on the graph.

$S_2 = P(1+i) + i[P(1+i)] = P(1+i)^2$ —(Second conversion period)
$S_3 = P(1+i)^2 + i[P(1+i)^2] = P(1+i)^3$ —(Third conversion period)
$S_4 = P(1+i)^3 + i[P(1+i)^3] = P(1+i)^4$ —(Fourth conversion period)

Through an inductive argument over n conversion periods, one would arrive at the fundamental single payment compound amount equation of:

$S_n = P(1+i)^{n-1} + i[P(1+i)^{n-1}] = P(1+i)^n$ —(nth conversion)

Therefore, it can be easily seen that the total compound amount always consists of a series of simple interest amounts based on the successive changes of principal incurred by the process of adding simple interest amounts at each conversion period.

Interest Rates

Generally, interest is expressed as a percent rate per year rather than a dollar amount per year. The dollar amounts of interest per year, however, are used for determining the interest rate per year.

To determine the percent rate per year, the dollars of interest earned is used as the numerator and either the principal P is used as the denominator, or the future sum S is used as the denominator. It, of course, should be remembered, that dollars of interest I divided by the principal P is the simple interest rate, while interest I divided by sum S is the simple discount rate. Thus, the base upon which interest rates are calculated becomes one of the important factors in understanding the mathematics of interest problems.

Nominal and Effective Rates

Two additional rates are of importance in financial usage; one is called the nominal rate of interest while the other is called the effective rate of interest.

The interest rate symbol i has been used previously as a percentage of the principal at each conversion period. A special case of this compound interest rate i is where the conversion period is one year. This special case of a compound interest rate is known as the nominal interest rate. The symbol for this rate is j.

Then,

$$j = im$$

where

> m is conversion periods per year

When the interest is converted into principal one or more times per year, the resulting annual interest rate is called the effective interest rate per year. When no conversions are incurred per year, the effective rate is equal to the nominal rate per year. Let j' represent the symbol for the effective interest rate. The effective interest in *Figure 2-2* is represented by line *QK*.

For example, if the nominal compound interest rate was 6 percent, then semiannual compounding ($m = 2$) would provide the amount at the end of one year of:

$$\left(1 + \frac{0.06}{2}\right)^2 = (1.03)^2 = 1.0609$$

in which $j' = 6.09$ percent effective interest in contrast to 6 percent nominal interest rate. More generally one could relate the nominal and effective compound interest rates as:

$$j' = \left(1 + \frac{j}{m}\right)^m - 1 = \left[\left(1 + \frac{j}{m}\right)^{m/j}\right]^j - 1 \quad \dots \dots \dots \dots (6)$$

Examination of the above equation will demonstrate that the difference between effective and nominal compound interest rates increases with m but an extension of the 6 percent example, shown in tabular form below, clearly demonstrates that the *rate of change* in the difference between effective and nominal rates decreases with m as shown in Table 2-1.

Table 2-1. Rate of Change in Difference Between Effective and Nominal Rates

m	Equation	Effective Interest Rate (j')	Time Interval of Conversion Period	($j' - j$) (%)
1	$(1 + 0.06/1)^1 - 1$	0.0600	Year	0.0
2	$(1 + 0.06/2)^2 - 1$	0.0609	Semiannual	0.09
4	$(1 + 0.06/4)^4 - 1$	0.06136	Quarter	0.136
12	$(1 + 0.06/12)^{12} - 1$	0.06168	Month	0.168
365	$(1 + 0.06/365)^{365} - 1$	0.07183	Day	0.183
24 x 365	$(1 + 0.06/8760)^{8760} - 1$	0.061836	Hour	0.1836

The j' column in the above table indicates an apparent convergence upon a maximum rate of compound interest as m becomes infinitely large. Of course, when m gets larger and larger, the conversion period becomes smaller and smaller down to the limit that interest is continuously generated and converted to principal. Examination of Equation 6, the expression on the right, will reveal that the portion inside the brackets approaches e as m increases to the limit, so that in continuous compounding the effective interest rate is:

$$j' = e^j - 1 \tag{7}$$

The amount of a loan in which interest for a nominal rate j is continuously compounded is:

$$S_n = P(1 + j')^n = P(1 + e^j - 1)^n = Pe^{nj} \tag{8}$$

Equivalence

The world *equivalence* is defined as "of equal worth." If we wish to compare money amounts at different time periods, it is customary to make equivalent conversions by the use of an interest rate. For example, if a bank charges 5 percent interest on a loan of $100 for one year, the amount paid to the bank is $105. These two amounts, $105 and $100, are equivalent to each other for the time period specified and the quoted interest rate. This can be expressed more precisely as follows:

SIMPLE AND COMPOUND INTEREST

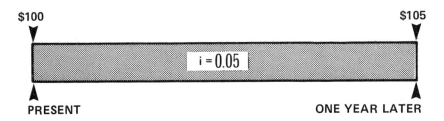

Fig. 2-3

This means that the $100 has been converted to an equivalent value one year later through the medium of an interest rate of 5 percent. It must be remembered, however, that money can have actual value at only one period of time, the present. Equivalence can be illustrated by the use of the following example problem:

Assume that a loan of $5000 is required by a borrower for a period of five years. The bank charges 5 percent. The bank advised the borrower that he can choose any of the following repayment plans:

(A) Pay $250 at the end of each of four years and $5250 at the end of the fifth year.

(B) Pay respectively $1250, $1200, $1150, $1100, and $1050 at the end of each of the five years.

(C) Pay $1155 at the end of each of the five years.

(D) Pay $6382 at the end of the fifth year.

The plans can be identified as:

(A-1) Principal amount is paid at end of term and interest due is paid annually.

(B-2) Principal is reduced by a constant annual payment and interest is paid on the balance of principal.

(C-3) Principal plus interest is paid by constant annual amounts.

(D-4) Principal plus interest accumulations are paid at the end of term.

These four plans are quantitatively developed in Table 2-2.

Table 2-2. Four Equivalent Repayment Plans (5 years @ 5%)

Plan	End of Year	Interest Due at Beginning of Year ($)	Total Money Due at End of Year ($)	Year End Payment ($)	Balance Due After Year End Payment ($)
A	0	—	—	—	5000
	1	250	5250	250	5000
	2	250	5250	250	5000
	3	250	5250	250	5000
	4	250	5250	250	5000
	5	250	5250	5250	—0—
		Sum of Year End Payments: $6250			
B	0	—	—	—	5000
	1	250	5250	1250	4000
	2	200	4200	1200	3000
	3	150	3150	1150	2000
	4	100	2100	1100	1000
	5	50	1050	1050	—0—
		Sum of Year End Payments: $5750			
C	0	—	—	—	5000
	1	250	5250	1155	4095
	2	205	4300	1155	3145
	3	157	3302	1155	2147
	4	108	2255	1155	1100
	5	55	1155	1155	—0—
		Sum of Year End Payments: $5775			
D	0	—	—	—0—	5000
	1	250	5250	—0—	5250
	2	263	5513	—0—	5513
	3	276	5789	—0—	5789
	4	289	6078	—0—	6382
	5	304	6382	6382	—0—
		Sum of Year End Payments: $6382			

SIMPLE AND COMPOUND INTEREST

Each of these plans is equivalent to $5000 and therefore equivalent to the others. For example, if the year end payments are discounted in Plan A to the present, the total present worth amounts over the five-year time period will be equal to $5000. Discounting the other plans in a similar manner will give a present worth of $5000 for each. Let us discount the year end payments in Plan A to show this basic relationship:

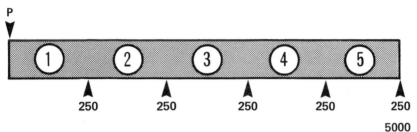

Fig. 2-4

Year	Payment at Year End	Discounted Value at P
1	$ 250	$ 238
2	250	227
3	250	216
4	250	205
5	250	196
	5000	3918

Sum of discounted amounts: $5000

The mathematical formulas commonly used to show equivalence between money amounts through the application of interest rates and time periods can be developed by the use of the following symbols:

i = Interest rate per period
F = Future sum of money
P = Present worth
n = Number of interest periods
A = Annuity constant

$$F = P(1+i)^n \quad \text{(Formula 1)}$$

$$P = \frac{i}{(1+i)^n} \quad \text{(Formula 2)}$$

$$A = F\frac{i}{(1+i)^n - 1} \quad \text{(Formula 3)}$$

$$A = P \frac{i(1+i)^n}{(1+i)^n - 1} \quad \text{................................} \quad \text{(Formula 4)}$$

$$F = A \frac{(1+i)^n - 1}{i} \quad \text{................................} \quad \text{(Formula 5)}$$

$$P = A \frac{(1+i)^n - 1}{i(1+i)^n} \quad \text{................................} \quad \text{(Formula 6)}$$

Formulas 1 and 2 are essentially single amount formulas. That is to say that a single amount is either compounded or discounted over a specified time period. Formulas 3, 4, 5, and 6 are essentially uniform series formulas. This uniformity is expressed by the annuity constant A in each equation. This means that A must be shown as some constant throughout each period in a time series. Several compound interest formula derivations and examples will serve to clarify the use of these equations:

Derivation of $F = P(1+i)^n$

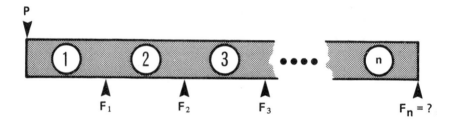

Fig. 2-5

$$F_1 = P + Pi = P(1+i)^1 \quad \text{................} \quad \text{(1st year conversion)}$$

$$F_2 = P + Pi + (P + Pi)i = P(1+i)^2 \quad \text{..........} \quad \text{(2nd year conversion)}$$

$$F_3 = P + Pi + (P + Pi)i + [P + Pi + (P + Pi)i]i$$
$$= P(1+i)^3 \quad \text{......................} \quad \text{(3rd year conversion)}$$

$$F_n = \ldots P(1+i)^n \quad \text{...................} \quad (n\text{th year conversion})$$

Note that for each year the process of compounding a single amount P is accomplished by adding the single amount P to the interest amount i for each of the time periods $1, 2, 3, \ldots, n$.

SIMPLE AND COMPOUND INTEREST

Example

What is the sum s for one, two, and three time periods when the beginning single amount P is equal to one dollar and the interest rate per period is 5 percent?

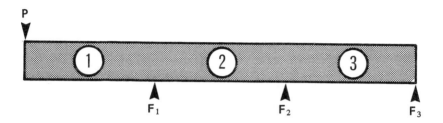

Fig. 2-6

Solution

$$F_1 = P(1+i)^1 = 1(1+0.05)^1 = \$1.0500$$

$$F_2 = P(1+i)^2 = 1(1+0.05)^2 = 1.1025$$

$$F_3 = P(1+i)^3 = 1(1+0.05)^3 = 1.1576$$

The effect of compounding can be seen easily by working the same problem using the simple interest formula:

$$F = P + Prt$$

$$F_1 = 1 + (1)(0.05)(1) = \$1.05$$

$$F_2 = 1 + (1)(0.05)(2) = 1.10$$

$$F_3 = 1 + (1)(0.05)(3) = 1.15$$

For the first period the simple and compound amounts are identical at $1.05 each. The second period calculations show a difference of $1.1025 - 1.1000 = 0.0025$. This is due to the compounding effect of 5 percent interest on the difference between $1.05 and $1.00. Thus, $(1.05 - 1.00)(0.05) = \$0.0025$. Similarly, the compound differences for the third period of $0.0076 would be: $0.05(1.1025 - 1.05) + 2(0.0025) = \$0.0076$.

The factor $(1+i)^n$ in the formula $F = P(1+i)^n$ is commonly called the future worth factor and its reciprocal $(1+i)^{-n}$ is commonly called the present worth factor.

Derivation of $F = A \left[\dfrac{(1+i)^n - 1}{i} \right]$

19

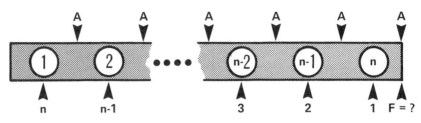

Fig. 2-7

This type of time series involving equal payments occurring at equal time intervals is sometimes called an annuity. This topic will be discussed more completely in a later chapter. The annuity formulas and factors are derived from "end of period" payments as shown in the above sketch. The F in the formula above is the sum of the compound amounts of each of the A payments. One important point to remember in deriving annuity formulas is the proper counting of time periods. This can be illustrated in Table 2-3, using F as the terminal point for reference.

Table 2-3. Calculation of Compound Amounts for Each R.

Payment at End of Time Period	Number of Time Periods Between Time of Payment and Terminal Point of Annuity	Compound Amount for Each R at the Terminal Point
1	$n-1$	$F(1+i)^{n-1}$
2	$n-2$	$F(1+i)^{n-2}$
....	....	....
$n-2$	2	$F(1+i)^2$
$n-1$	1	$F(1+i)^1$
n	0	$F(1+i)^0$

The sum of the compound amounts of each A payment is equal to F. Hence:

$$F = A\,[1 + (1+i)^1 + (1+i)^2 + (1+i)^3 \ldots + (1+i)^{n-2} + (1+i)^{n-1}]$$

Multiplying through by the common factor $(1+i)$, the preceding equation becomes:

$$F(1+i) = A\,[(1+i)^1 + (1+i)^2 + (1+i)^3 + (1+i)^4 \ldots + (1+i)^{n-1} + (1+i)^n]$$

Subtracting one from the other:

$$F = A\left[\frac{(1+i)^n - 1}{i}\right]$$

SIMPLE AND COMPOUND INTEREST

This equation is used to find the compound sum of a series of constant payments when A, n and i are known. The F equations is used to find the sum of the compound amounts F when A, n and i are known.

Example:

What is the sum of a series of year end payments of $100 each for a term of five years at 5 percent payable annually? ($i = 0.05$; $n = 5$)

The sketch of the time series for this example would be:

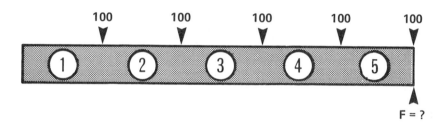

Fig. 2-8

Solution:

$$F = 100 \times \frac{(1 + 0.05)^5 - 1}{0.05}$$

$$= 100 \times \frac{1.276275 - 1}{0.05} = \$100 \ (5.5255)$$

$$= 552.55$$

Table 2-4 shows the detailed operation of the annuity plan.

This table show that payment No. 1 is compounded four periods in the term; No. 2, three periods; No. 3, two periods, No. 4, one period, and No. 5, zero periods. Thus, if an amount A is paid at the end of each period for a total of n periods, with each A accumulating at an interest rate of i per period, then the sum F is equal to the sum of all constant payments A plus accumulated interest for the length of the term.

This same formula can be used to find the constant amount A by transforming to:

$$A = F \left[\frac{1}{(1 + i)^n - 1} \right]$$

ECONOMICS OF MACHINE TOOL PROCUREMENT

Table 2-4. End of Period Amounts for Each Year

End of Year Period	Sum at Beginning of Each Period $	Total Interest Earned at End of Each Period $	Constant Amount Paid at End of Each Period $	Total Sum at End of Each Period $
1	0.0	0.0	100	100.00
2	100.00	5.00	100	205.00
3	205.00	10.25	100	315.25
4	315.25	15.76	100	431.01
5	431.01	21.55	100	552.56

In this form, the factor is commonly called the sinking fund factor and is used principally to accumulate some fixed amount over a stated time period at some stated interest rate.

Example:

An individual wishes to accumulate a sum of $500 at the end of five years with an interest rate of 5 percent. What must the constant deposit be at the end of year? ($i = 0.05$; $n = 5$) First, set up the time series as follows:

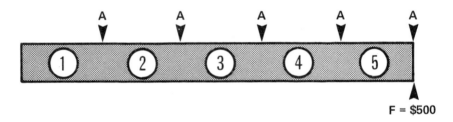

F = $500

Fig. 2-9

$$A = \$500 \left[\frac{0.05}{(1+0.05)^5 - 1} \right]$$

$$A = \$500 \left[\frac{0.05}{1.276275 - 1} \right]$$

$$= \$90.485$$

The tabulation, Table 2-5, shows the detailed operation of the annuity plan to accumulate a sinking fund of $500.

SIMPLE AND COMPOUND INTEREST

Table 2-5. Annuity Plan to Accumulate a Sinking Fund of $500

End of Year Period	Sum at Beginning of Each Period $	Total Interest Earned at End of Each Period $	Constant Amount Paid at End of Each Period $	Total Sum at End of Each Period $
1	0.0	0.00	90.485	90.485
2	90.485	4.524	90.485	185.494
3	185.494	9.275	90.485	285.254
4	285.254	14.265	90.485	390.002
5	390.002	19.500	90.485	500.000

This type of plan, although an annuity, is commonly called a sinking fund. If an obligation due at some future date is met by depositing or investing equal sums at equal time intervals, the total sum in the sinking fund at any date is the amount formed by payments into the fund plus the compound interest earned as of that date.

Derivation of $P = A \dfrac{(1 + i)^n - 1}{i(1 + i)^n}$

The factor in this formula is commonly called the annuity present worth factor. The A indicates that an annuity is represented and that the A is a constant amount. The P represents the sum of all the discounted A's present in the term. This means that each A is discounted each period (compounding the discount) until the present worth P is reached. The formula can be derived as follows:

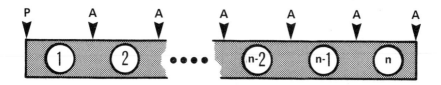

Fig. 2-10

Tabulation of this time series is shown in Table 2-6.

23

ECONOMICS OF MACHINE TOOL PROCUREMENT

Table 2-6. Calculation of Discounted Amount for A at Beginning Date of Annuity

Payment at End of Time Period	Number of Time Periods Between Time of Payment to Beginning Date of Annuity	Discounted Amount for Each A at Beginning Date of Annuity
1	1	$A(1+i)^{-1}$
2	2	$A(1-i)^{-2}$
$n-2$	$n-2$	$A(1+i)^{-(n+2)}$
$n-1$	$n-1$	$A(1+i)^{-(n+1)}$
n	n	$A(1+i)^{-n}$

The sum of the discounted amounts of each A is equal to the present worth P. Hence

$$P = A\,[(1+i)^{-1} + (1+i)^{-2} \ldots + (1+i)^{-n+2} + (1+i)^{-n+1} + (1+i)^{-n}] \qquad (1)$$

Multiply through by $(1+i)^{-1}$

$$P(1+i)^{-1} = A\,[(1+i)^{-2} + (1+i)^{-3} \ldots + (1+i)^{-n+1} + (1+i)^{-n} + (1+i)^{-n-1}] \qquad (2)$$

Subtracting the first equation from the second equation,

$$P\,[(1+i)^{-1} - 1] = A\,[(1+i)^{-n-1} - (1+i)^{-1}] \qquad (3)$$

Convert $(1+i)^{-1} - 1$ to $\dfrac{-i}{(1+i)}$

This can be accomplished as follows:

Multiply and divide the expression by $(1+i)$

Then $(1+i)^{-1} - 1 = \dfrac{(1+i)^0 - (1+i)}{(1+i)}$

$\dfrac{1-1-i}{(1+i)} = \dfrac{-i}{(1+i)}$

Then equation (3) becomes,

$$P\left[\dfrac{-i}{(1+i)}\right] = A\left[(1+i)^{-n-1} - (1+i)^{-1}\right] \qquad (4)$$

24

SIMPLE AND COMPOUND INTEREST

Multiply both sides of equation (4) by $-(1 + i)$. Then

$$P\left[\frac{i(1+i)}{(1+i)}\right] = A(1+i)(1+i)^{-1} - (1+i)(1+i)^{-n-1}$$

$$Pi = \left[A\ 1 - (1+i)^{-n}\right]$$

Multiply numerator and denominator of equation by $(1 + i)^n$. Then

$$Pi = A\left[\frac{(1+i)^n - 1}{(1+i)^n}\right]$$

$$P = A\left[\frac{(1+i)^n - 1}{i(1+i)^n}\right]$$

In this form the factor commonly is called the uniform present worth factor and is used to convert a series of constant A's to a present worth amount.

Example:

If money is worth 5 percent compounded annually, how much money is required now to purchase an annuity of $100 per year for five years?

$$P = A\left[\frac{(1+i)^n - 1}{i(1+i)^n}\right] = 100\left[\frac{(1.05)^5 - 1}{0.05(1.05)^5}\right]$$

$$= 100\left[\frac{1.276275 - 1}{0.05(1.276275)}\right] = 100(4.3294)$$

$$= \$432.94$$

The tabulation in Table 2-7 shows the detailed operations of the annuity plan.

The discounted amounts in Column 4 are equal to the sum of the constant payment and the total present worth for the beginning of the year indicated minus total present worth for end of year indicated. Thus, $100.00 + 95.24 − 185.94 = $9.30 for the second year, etc.

ECONOMICS OF MACHINE TOOL PROCUREMENT

Table 2-7. Annuity Value for $100 Annually at 5%

Year	Total Present Worth For Beginning of Year Indicated $	Constant Payment For End of Year Indicated $	Amount of Discount for Year Indicated $	Total Present Worth For End of Year Indicated $
1	0.00	100.00	4.76	95.24
2	95.24	100.00	9.30	185.94
3	185.94	100.00	13.62	272.32
4	272.32	100.00	17.73	354.59
5	354.59	100.00	21.65	432.94

The time series for this tabulation would be:

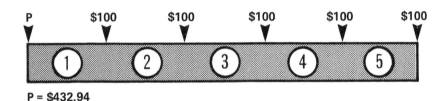

Fig. 2-11

The first A of $100 is discounted for one period, the second A of $100 is discounted for two periods, etc. The sum of the discounted amounts for each A of $100 is, of course, equal to the present worth of the series, represented by the letter P.

The uniform present worth equation can be converted to

$$A = P\left[\frac{i(1+i)^n}{(1+i)^n - 1}\right]$$

In this form, the factor commonly is called the uniform capital recovery factor. It is used principally to recover an investment through constant periodic payments. Each constant payment consists of a portion of the present worth or principal P and a portion of interest the sum total of which is equal to the constant A.

Example

If money is worth 4 percent compounded annually each year for five years, what constant amount will $500 buy now?

SIMPLE AND COMPOUND INTEREST

$$A = P\left[\frac{i(1+i)^n}{(1+i)^n-1}\right] = \frac{(500)(0.05)(1+0.05)^5}{(1+0.05)^5-1}$$

$$= \frac{(500)(0.05)(1.276275)}{1.276275-1}$$

$$= \$115.49$$

The sketch of the time series for this operation would be:

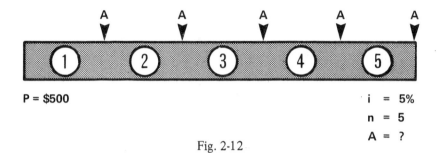

P = $500   i = 5%
n = 5
A = ?

Fig. 2-12

The factors used in this chapter to solve equivalence problems were calculated directly from the data given in the various problems. The following symbols are suggested for identification of the interest factors and the formula to which it applies. The script letter "g" means *given*; the script letter "f" means *find*.

	Formula	Factor of Identification
1)	$F = P(1+i)^n$	$(gPfF)^i_n$
2)	$P = \dfrac{F}{(1+i)^n}$	$(gFfP)^i_n$
3)	$A = F \dfrac{i}{(1+i)^n-1}$	$(gFfA)^i_n$
4)	$F = A \dfrac{(1+i)^n-1}{i}$	$(gAfF)^i_n$
5)	$A = P \dfrac{i(1+i)^n}{(1\ i)^n-1}$	$(gPfA)^i_n$
6)	$P = A \dfrac{(1+i)^n-1}{i(1+i)^n}$	$(gAfP)^i_n$

The use of the interest factors and the identification symbols can be illustrated as follows:

Examples:

What is the future amount paid on a loan of $1000 with interest compounded yearly at 5 percent for five years.

Solution:

$$F = P(1+i)^n = 1000 \; (gPfF)^{0.05}_{5} = \$1000 \; (1.276) = \$1276$$

Example:

Find the amount that would be accumulated through depositing $50 in a fund every year for ten years. Interest is 4 percent and is compounded yearly.

Solution:

$$F = A \left[\frac{(1+i)^n - 1}{i} \right] = 50 \; (gAfF)^{0.04}_{10} = 50 \; (12.006)$$
$$= \$600.30$$

CHAPTER 3

THE DEPRECIATION METHOD DECISION

Earlier discussions have shown that a necessary cost of production is the loss in value of assets necessary for production. This cost is known as the depreciation charge. However, it is extremely difficult to precisely estimate an asset's loss of value over a period of time. Even if one can do so for one asset, the loss for another asset may follow entirely different patterns. This dilemma would prove to be an impossible accounting task in determining the ownership of the firm unless some arbitrary but systematic procedure is used. Such procedures are defined as *accounting depreciation methods*, the subject of this chapter. The decision faced by management is in determining which method to employ. Methods which tend to give a more realistic portrayal of market value provide better estimates of ownership but often involve greater accounting costs.

Another feature of the depreciation decision is the effect of the choice on profit-based taxes. The relationship between depreciation and taxes occurs because depreciation is an allowable expense to the firm, and added allowable expenses decrease net profit. With the sliding tax rate, a depressed net profit also decreases the tax rate so that the entire tax cost is reduced. However, this effect is more illusory than real because higher depreciation charges early in an asset's life means lower charges later. Nevertheless, there are some implications of the depreciation decision on taxes later in this chapter.

This depreciation method decision may be viewed as shown in *Figure* 3-1. The criterion of choice may be the lowest accounting cost, the greatest accounting information, the highest rate-of-return resulting or some combination of these possible criteria.*

Depreciation Methods

The depreciation method was defined earlier as the rule which determined the amount of each depreciation charge over the service period in which an asset is depreciated. This rule may create a schedule of charges which are uniform, or increase or decrease over the service life as long as the depreciation fund at the end of the service life exactly equals the initial cost of the asset less its expected salvage value. In the discussion to follow, various methods of depreciation will be discussed with means of calculating depreciation charges, the value of the depreciation fund, and the book value.

Straight-line Depreciation

The straight-line depreciation method (SL) is defined as a uniform series of

---

*See B. Alva Schoomer, Jr., "Optimal Depreciation Strategies for Income Tax Purposes," *Man. Sci.*, Vol. 12, No. 12, August 1966, pp. B-552-B-579, for extensions of this concept of selecting depreciation methods.

ECONOMICS OF MACHINE TOOL PROCUREMENT

Figure 3-1. Depreciation Method Decision Example in Standard Format

Method	Future
a_1 : Straight Line	Earnings per year on time–value of funds held less cost of employing the method.
a_2 : Percentage of Declining Balance	Earnings per year on time–value of funds held less cost of employing the method.
a_3 : Sum of Digits	Earnings per year on time–value of funds held less cost of employing the method.
a_4 : Double Declining	Earnings per year on time–value of funds held less cost of employing the method.
a_5 : Sinking Fund	Earnings per year on time–value of funds held less cost of employing the method.

depreciation charges over the service period of N years in which each depreciation charge is exactly $1/N$th of the difference between the initial cost and the salvage value. The depreciation charge in the year t' for $0 \leq t' \leq N$ is:

$$f(t') = \frac{I - L}{N}$$

where t' = year, I = original investment, L = salvage value, and $f(t')$ = depreciation charge.

It then follows that the depreciation fund, $F(t)$, receives this charge each year, so that by the t th year the fund builds to:

$$F(t) = \sum_{t'=1}^{t} \frac{I-L}{N} = \frac{t}{N}(I-L)$$

and the book value for the t th year is:

$$I - F(t) = I - \frac{t}{N}(I - L)$$

As an example, a $1000 machine with zero salvage value at the end of a ten-year life would have a $100 depreciation charge each year if the straight-line

THE DEPRECIATION METHOD DECISION

method is used. After six years the depreciation fund would have accumulated to

6/10 ($1000 − 0)= $600 and the book value would be
$1000 − $600 = $400.

Clearly the straight-line depreciation method provides computational simplicity, which accounts for its wide-spread use.* However, the SL suffers from two disadvantages:
1. There is a tax disadvantage associated with it
2. It does not provide a good approximation of value depreciation for most capital equipment.

An example of the use of the SL depreciation method may be provided by assuming that the method applies to a machine with a $100 first cost, a $10 salvage value and a nine-year service period. In this situation the funds to be recovered over the service period are:

$$I - L = 100 - 10 = 90$$

and the uniform depreciation charge needed to recover the funds is:

$$f(t') = \frac{I-L}{N} = \frac{90}{9} = 10$$

Consequently, the schedule of depreciation charges, fund and book value for this example would appear as shown in Table 3-1.

The Fixed Percentage of a Declining Balance Method

Fixed percent methods, declining balance methods, and the fixed percentage of a declining balance method (FPDB) are various names for a depreciation method which applies a constant percentage r to the previous year's book value to get the current year's depreciation charge. Since each year's book value decreases, a common percentage of decreasing values provides a decreasing schedule of depreciation charges. In the first year the depreciation charge is determined by applying the fixed percentage r against the book value in the year zero, or:

$$f(1) = r(I)$$

In the second year the depreciation charge is:

$$f(2) = r[I - f(1)]$$
$$= r(I - rI)$$
$$= rI(1-r)$$

---

*The Internal Revenue Service also encouraged the use of the straight-line depreciation method for many years, due to the tax laws and auditing convenience.

Table 3-1. Example of the Straight-line Depreciation Method

Year	Depreciation Charge	Depreciation Fund	Book Value
t or t'	$f(t')$	$F(t)$	$I - F(t)$
1	10.00	10.00	90.00
2	10.00	20.00	80.00
3	10.00	30.00	70.00
4	10.00	40.00	60.00
5	10.00	50.00	50.00
6	10.00	60.00	40.00
7	10.00	70.00	30.00
8	10.00	80.00	20.00
9	10.00	90.00	10.00
Sum	90.00	--	--

It thus follows that the depreciation charge in the tth year is:

$$f(t) = r[I - F(t-1)]$$
$$= rI(1-r)^{t-1}$$

In the last year of the service period the depreciation charge is:

$$f(N) = r[I - F(N-1)]$$
$$= rI(1-r)^{N-1}$$

The book value of the equipment decreases over the service period. At the end of the first year a depreciation charge of rI dollars is made into the depreciation fund so that the book value is:

$$I - F(1) = I - rI$$
$$= I(1-r)$$

By the end of the second year, the depreciation charges levied build the fund to:

$$F(2) = rI + rI(1-r)$$
$$= rI[1 + (1-r)]$$

Consequently the book value is:

$$I - F(2) = I - rI [1 - (1-r)]$$
$$= I [1 - 2r + r^2]$$
$$= I(1-r)^2$$

It follows through similar reasoning that the book value in the tth year is:

$$I - F(t) = I(1-r)^t$$

The book value in the Nth or last year follows directly by substitution, but also it is recalled that the book value in the final year must equal the salvage value as:

$$I(1-r)^N = L$$

Since the initial value I, salvage L and the service period N are assumed to be fixed, it follows that there is one—and only one—value of the fixed percentage r which will make the book value in the final year equal the salvage value, and that is:

$$r = 1 - \sqrt[N]{L/I} \text{ for } L > 0$$

Because the value of r is undefined for a zero salvage value in this equation, it is convenient to use the following technique: Assume a nominally small salvage value, calculate r and apply it in the manner prescribed for every year except the final year. In the final year assume a straight-line depreciation charge in order to arrive at a final book value of zero. The regular charge, fund, and book values are summarized in Table 3-2 for the Fixed Percentage Declining Balance depreciation method (FPDP).

Table 3-2. **Summary of Calculation Expressions for the Fixed Percentage of a Declining Balance Depreciation Method**

Year t or t'	Depreciation Charge $f(t')$	Depreciation Fund $F(t)$	Book Value $I - F(t)$
1	rI	rI	$I(1-r)$
2	$rI(1-r)$	$rI[1+(1-r)]$	$I(1-r)^2$
3	$rI(1-r)^2$	$I[1-(1-r)^3]$	$I(1-r)^3$
t	$rI(1-r)^{t-1}$	$I[1-(1-r)^t]$	$I(1-r)^t$
N	$rI(1-r)^{N-1}$	$I[1-(1-r)^N] = I - S$	$I(1-r)^N = L$

In practice, the choice of a percentage differs from the academic formula for the purpose of computation simplicity and because of tax ruling limitations.* Although the calculations can be slightly simplified by selecting a rounded value for the fixed percentage, the resulting calculations are still complex relative to the straight-line method. This complexity is largely responsible for the small usage of the FPDP method. Despite its low usage, this method is presented here because its book value exemplifies the value depreciation of most types of equipment. Thus, its use provides accounting with a better approximation of realizable market value. Its use also improves the information which accounting provides.

Let us consider the FPDP depreciation method with the example used with the SL method. The calculation of the fixed percentage is:†

$$r = 1 - \sqrt[N]{L/I}$$
$$= 1 - \sqrt[9]{10/100}$$
$$= 1 - 0.7743$$
$$= 0.2257$$

Table 3-3. Example of the Fixed Percentage of a Declining Balance Depreciation Method

Year t or t'	Depreciation Charge $f(t')$	Depreciation Fund $F(t)$	Book Value $I - F(t)$
1	22.57	22.57	77.43
2	17.48	40.05	59.95
3	13.53	53.58	46.42
4	10.48	64.06	35.94
5	8.11	72.17	27.83
6	6.28	78.45	21.55
7	4.86	83.31	16.69
8	3.77	87.08	12.92
9	2.92	90.00	10.00
Sum	90.00	--	--

*These rulings vary from time to time and from agency to agency, but one example of a limitation is the federal regulation which does not allow a depreciation rate greater than twice that determined by the straight-line depreciation method.

†Logarithm tables or log-log slide rule scales are frequently very useful in the calculation of the fixed percentage.

Using this calculated value in the expressions summarized in Table 3-2, these calculations result in the depreciation schedule shown in Table 3-3.*

Sum-of-the-Years'-Digits Depreciation Method

A method of depreciation which eliminates most of the calculation disadvantages of the fixed percentage method but retains the advantage of resembling value depreciation is called the sum-of-the-years'-digits method or SYD. Generally, this method gives a relatively fast writeoff in the early years, a relatively slow writeoff in the late years, and a complete writeoff of the depreciable amount over the estimated life.

The sum of years digits refers to the sum total of the years of estimated life of the asset or the service period. Thus, if the estimated life of a machine is nine years, the sum of digits would be:

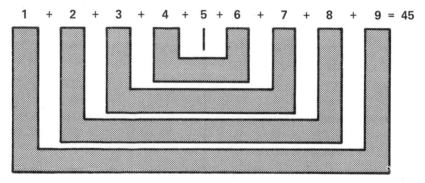

Fig. 3-1

For long service periods the calculation of this digits' sum can be shortened by observing that the pairs of digits all sum to $N + 1$ and that with $N/2$ of these subsums, the total sum of the digits is:

Sum of Digits = $\dfrac{N(N+1)}{2}$

In the example given, the equation yields,

$\dfrac{9(10)}{2} = \dfrac{90}{2}$

$= 45$

---

*The reader may have observed that the FPDB method's first depreciation charge is more than twice that of the SL method. The Internal Revenue Service would not permit this practice according to current rulings.

This sum of digits is used in this depreciation method to form the denominator of a depreciation charge factor. The numerator of this factor is formed by the remaining years in the service period. A simple way of representing the numerator values is to reverse the order of digits shown in Table 3-4.

Table 3-4. Calculating Depreciation Factor for SYD Depreciation Method

										Sum
Year	1	2	3	4	5	6	7	8	9	45
Reverse Order	9	8	7	6	5	4	3	2	1	45
Depreciation Factor	$\frac{9}{45}$	$\frac{8}{45}$	$\frac{7}{45}$	$\frac{6}{45}$	$\frac{5}{45}$	$\frac{4}{45}$	$\frac{3}{45}$	$\frac{2}{45}$	$\frac{1}{45}$	$\frac{45}{45}$

The depreciation charge $f(t')$ for the sum-of-years'-digits method can be developed into a general formula by deductive logic as follows: Let I = initial investment, L = salvage value, N = years of estimated life, and t' = any year.

Then,

$$f(1) = \frac{N}{N(N+1)/2}(I-L) = \frac{2}{N+1}(I-L)$$

$$f(2) = \frac{N-1}{N(N+1)/2}(I-L) = \frac{2(N-1)}{N(N+1)}(I-L)$$

$$f(3) = \frac{N-2}{N(N+1)/2}(I-L) = \frac{2(N-2)}{N(N+1)}(I-L)$$

$$f(4) = \frac{N-3}{N(N+1)/2}(I-L) = \frac{2(N-3)}{N(N+1)}(I-L)$$

$$f(t') = \frac{N-(t'-1)}{N(N+1)/2}(I-L) = \frac{2(N+1-t')}{N(N+1)}(I-L)$$

By applying this general equation $f(t')$ to any year, the following can be developed.

(Year 1) $\qquad f(t') = \frac{2(N+1-t')}{N(N+1)}(I-L)$

THE DEPRECIATION METHOD DECISION

$$f(1) = \frac{2(N+1-1)}{N(N+1)}(I-L)$$

$$= \frac{2N}{N(N+1)}(I-L)$$

(Year 4) $\quad f(t') = \frac{2(N+1-t')}{N(N+1)}(I-L)$

$$f(4) = \frac{2(N+1-4)}{N(N+1)}(I-L)$$

$$= \frac{2(N-3)}{N(N+1)}(I-L)$$

In the general equation

$$f(t') = \frac{2(N+1-t')}{N(N+1)}(I-L)$$

the depreciation charge factor multiplied by the depreciable amount $(I-L)$ is equal to the depreciation charge.

The depreciation charge factor in the general equation is equal to

$$\frac{2(N+1-t')}{N(N+1)}(I-L)$$

This factor can be rewritten as follows:

$$2(N-t'+1) \div \sum_{t'=1}^{N} t'$$

Thus for an estimated life of nine years,

$$\sum_{t'=1}^{N} t' = 1+2+3+4+5+6+7+8+9 = 45$$

and $N - t' + 1$ for nine years is equal to

$$\frac{9-9+1}{45} = \frac{1}{45}$$

so that for the ninth year the depreciation charge factor is 1/45.

Thus, the depreciation factor for the sixth year may be obtained by formula as:

$$\text{Depreciation charge factor } \frac{9-6+1}{45} = \frac{4}{45}$$

$$\text{Depreciation charge } f(t') = \frac{2(N-t'+1)}{N(N+1)}(I-L)$$

$$= \frac{2(9-6+1)}{9(10)}(100-10)$$

$$= \frac{4}{45}(90)$$

$$= 8$$

or through the reverse order technique shown in Table 3-4; the depreciation charge is simply the product of the depreciable amount and the depreciation factor. Consequently, the changes in depreciation charges each year depend only upon the length of the service period and not upon the initial or salvage values; all that is needed for calculation is a table of factors for each different service period. Thus, the depreciation fund is:

$$F(t) = \sum_{t'=1}^{t} f(t') = \sum_{t'=1}^{t} \frac{N-t'+1}{N(N+1)} \times (I-L)$$

which simplifies to:

$$F(t) = \left(\frac{2N-t+1}{N+1}\right)\left(\frac{t}{N}\right) \times (I-L)$$

These formulas are summarized in Table 3-5. The supplementary calculations are given in Table 3-6.

Let us view the same illustrating depreciation problem to describe the SYD depreciation method. In this example the service period is nine years so that the denominator to the depreciation charge factor is:

$$\frac{N(N+1)}{2} = \frac{9(10)}{2} = 45$$

38

THE DEPRECIATION METHOD DECISION

Table 3-5. Summary of Calculation Expressions for the Sum-of-the-Years'-Digits Depreciation Method

Year	Depreciation Charge	Depreciation Fund	Book Value
t or t'	$f(t')$	$F(t)$	$I - F(t)$
1	$\dfrac{2}{N+1}(I-L)$	$\dfrac{2}{N+1}(I-S)$	$I - \dfrac{2}{N+1}(I-L)$
2	$\dfrac{2(N-1)}{N(N+1)}(I-L)$	$\dfrac{(4N-2)}{N(N+1)}(I-L)$	$I - \dfrac{(4N-2)}{N(N+1)}(I-L)$
t	$\dfrac{2(N+1-t)}{N(N+1)}(I-L)$	$\dfrac{(2N-t+1)}{(N+1)}\left(\dfrac{t}{N}\right)(I-L)$	$I - \dfrac{(2N-t+1)}{(N+1)}\left(\dfrac{t}{N}\right)(I-L)$
N	$\dfrac{2(N+1-N)}{N(N+1)}(I-L)$	$(I-L)$	$I - (I-L) = L$

Table 3-6. Example of the Sum-of-the-Years'-Digits Depreciation Method

Year	Reverse	Depreciation Factor	Depreciation Charge	Depreciation Fund	Book Value
t or t'	$N - t' + 1$	$\left(\dfrac{N-t'+1}{45}\right)$	$f(t')$	$F(t)$	$I - F(t)$
1	9	9/45	18.00	18.00	82.00
2	8	8/45	16.00	34.00	66.00
3	7	7/45	14.00	48.00	52.00
4	6	6/45	12.00	60.00	40.00
5	5	5/45	10.00	70.00	30.00
6	4	4/45	8.00	78.00	22.00
7	3	3/45	6.00	84.00	16.00
8	2	2/45	4.00	88.00	12.00
9	1	1/45	2.00	90.00	10.00
Sum	45	45/45	90.00	--	--

Double Declining Balance Depreciation Method

After 1954, the tax laws governing depreciation methods were revised and a new ruling came into being which allowed any reasonable depreciation so long as the depreciation charge in any year did not exceed twice that of the SL method. A result of this new ruling is the double declining balance or DDB depreciation method.

This method achieves the maximum writeoff speed in the first year of two times the straight-line depreciation charge or:

$$F(1) = 2\frac{I-L}{N} = \frac{2}{N}(I-L)$$

A more typical approach with this method is to consider double the percentage of the depreciation amount per year for the SL method, otherwise known as the depreciation rate, as:

$$2\left(\frac{100\%}{N}\right) = \frac{2}{N}100\% = r'$$

The depreciation charge in the first year is:

$$f(1) = r'(I-L)$$

and thereafter the r' percentage is applied to the book value from the previous year, as shown for the fixed percentage of a declining balance method, except for the straight-line runout in the final year. Consequently, the calculation expressions in Table 3-2 may be used with r' in lieu of r with the exception of the final year. In the final year, or for that matter, any year desired as a switch point, the remaining depreciation may be changed to a SL method over the remaining portion of the service period. For a switch point in the final year of the DDB method, the final depreciation charge is equal to the difference between the book and salvage values:

$$F(N) = I(1-r^1)^{N-1} - L$$

This switch over to the SL method is absolutely necessary unless r' happens to equal exactly the academic fixed percentage r.

Applying the DDB method to the illustration example for depreciation methods involves first the calculation of r' as:

$$r' = \frac{2}{N} = \frac{2}{9} = 0.2222$$

and then the remaining calculations shown in Table 3-7.

Table 3-7. Example of the Double Declining Balance Depreciation Method

Year t or t'	Depreciation Charge $r'[I - F(t-1)]$	Depreciation Fund $F(t)$	Book Value $I - F(t)$
0	0	0	100.00
1	22.22	22.22	77.78
2	17.28	39.50	60.50
3	13.44	52.94	47.06
4	10.46	63.40	36.60
5	8.13	71.53	28.47
6	6.33	77.86	22.14
7	4.92	82.78	17.22
8	3.83	86.16	13.39
9	3.39	90.00	10.00
Sum	90.00	— —	— —

Switch-over Depreciation Methods

When some curvilinear depreciation is used, such as the fixed percentage of a declining balance, SYD or DDB methods, it is common practice to change over to the straight-line method after the service period's midpoint. The year in which the change over occurs is often called the switch-over point.

An example of a combined method is portrayed by using a five-year switch-over point for the SYD method in this continuing depreciation example in Table 3-8.

Sinking Fund Depreciation Method

The sinking fund or SF method is seldom used today except by nonprofit institutions. It does have an advantage not common to the other depreciation methods, however, in that it recognizes the time value of money. As the name implies, a fund is established by depositing or investing equal sums at equal time intervals. The fund so established accrues interest at some specified rate so that at the end of the estimated service life, the sum total of all deposits to the fund *plus the accrued interest* is equal to the total depreciable amount $I - L$. The depreciation charge at the end of any time period within the service life is equal to the total amount then in the fund.

It should be understood that the major purpose of a sinking fund is to retire an indebtedness, including the interest thereon at the rate specified in the terms

Table 3-8. A Switch-over SYD Depreciation Method Example

Year t or t'	Depreciation Charge $f(t')$	Depreciation Fund $F(t)$	Book Value $I - F(t)$
1	18.00	18.00	82.00
2	16.00	34.00	66.00
3	14.00	48.00	52.00
4	12.00	60.00	40.00
5	10.00	70.00	30.00
6	5.00	75.00	25.00
7	5.00	80.00	20.00
8	5.00	85.00	15.00
9	5.00	90.00	10.00
Sum	90.00	--	--

of the obligation. From a depreciation viewpoint, this means that in our machine example of a $100 first-cost, nine-year service period, and $10 salvage, the sinking fund at the end of nine years would accrue to exactly $90 including interest earned on the fund at some specified rate.

Traditionally, sinking fund interest rates are somewhat low, at 3, 4, or 6%. This would be a good reason for a company not to establish a sinking fund for depreciation recovery, when, let us say, the company is earning 15 to 20% on its investment before taxes. However, in most cases where the sinking fund method is used, a hypothetical fund is established and only rarely would an actual fund be used. Thus, the important features of the sinking fund method resolve into the mathematics involved rather than terminology.

There is one serious disadvantage to this plan. The mathematics of the plan indicate that a larger depreciation amount be charged in the later years of the asset's life and a correspondingly smaller depreciation amount charged in the early years. This means that a relatively large book value is carried in the accounts during the early years. If the asset is sold for some reason during the early years, the difference between the book value and the realizable value may be sizable. Therefore, the sinking fund method does not reflect the fact that most assets lose value most rapidly in the early years of life. The depreciation example with the sinking fund method is shown in Table 3-9 to illustrate these points.

To determine the depreciation charge at 6% for the sinking fund method, the 6% sinking fund factor-uniform series must be used. This factor is determined by using compound interest tables. The sinking fund factor for nine years and 6% is

THE DEPRECIATION METHOD DECISION

0.08702. The annual cost A to the sinking fund is:

$$A = F \text{ (Sinking Fund Factor)}$$
$$= 90 (0.08702)$$
$$= \$7.83/\text{year}$$

This amount does not represent the depreciation because interest must be added each year to derive the depreciation charge. This charge is variable and can be derived by use of a simple table and time series.

Table 3-9. Example of the Sinking Fund Depreciation Method

Year t or t'	Annual Cost	$f(t)$*	Depreciation Fund $F(t)$	Book Value $I - F(t)$
1	7.83	7.83	7.83	92.17
2	7.83	8.30	16.13	83.87
3	7.83	8.80	24.93	75.07
4	7.83	9.33	34.26	65.74
5	7.83	9.89	44.15	55.85
6	7.83	10.48	54.63	45.37
7	7.83	11.11	65.74	34.26
8	7.83	11.78	77.52	22.48
9	7.83	12.48	90.00	10.00
Sum	70.47	90.00	--	--

*Depreciation charge at 6% interest

In this example problem, the amount of $7.83 is deposited each year into the 6% sinking fund and each deposit draws interest for a variable time.

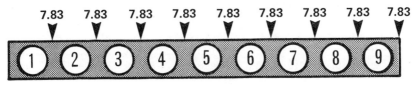

TIME SERIES

Fig. 3-2

Each deposit is made at the end of each year including one at the end of the ninth year. Thus, the first deposit draws compound interest for eight years, and the last deposit draws zero interest. The total difference in the fund between successive years then becomes the depreciation charge for each year. The sum of the yearly charges must, in this example, amount to $90. This can be clarified by use of the following table:

Chapter 3 Table (a)

End of Year	S at Beginning	Interest 6%	A Factor	S at End	F at End Column Differences
	a	b = (a × 0.06)	c	d = a + b + c	
1	0.00	0.0000	7.83	7.83	7.83 − 0.00 = 7.83
2	7.83	0.4698	7.83	16.13	16.13 − 7.83 = 8.30
3	16.13	0.9678	7.83	24.93	24.93 − 16.13 = 8.80
4	24.93	1.4958	7.83	34.26	34.26 − 24.93 = 9.33
5	34.26	2.0556	7.83	44.15	44.15 − 34.26 = 9.89
6	44.15	2.6480	7.83	54.63	54.63 − 44.15 = 10.48
7	54.63	3.2778	7.83	65.74	65.74 − 54.63 = 11.11
8	65.74	3.9444	7.83	77.52	77.52 − 65.74 = 11.78
9	77.52	4.6512	7.83	90.00	90.00 − 77.52 = 12.48

Total = 90.00

A more direct way to develop the sinking fund depreciation charge is by use of the single-payment compound amount factor as shown on the following page.

Service Output

Many other forms of depreciation methods have been suggested from time to time but most are seldom used today.* An exception is the so called *service-output* method. Essentially, this method uses the estimated total output from the equipment over an estimated service life; depreciation charges are made proportional to the ratio of actual output to total expected output. Of course, a close approximation to output is machine hours, so that the actual machine

---

*E. Paul DeGarmo, *Engineering Economy*, MacMillan Company, 1960, pp. 106-109, discusses the Gillette Formula and the present-worth depreciation formula as additional depreciation methods.

Chapter 3 Table (b)

End of Year		Single Payment Amount Factor	Sinking Fund Deposit		Sinking Fund Depreciation Charge
1	1 =	1.0000 X	7.83	=	7.83
2	$(1 + 0.06)^1$ =	1.0600 X	7.83	=	8.30
3	$(1 + 0.06)^2$ =	1.1236 X	7.83	=	8.80
4	$(1 + 0.06)^3$ =	1.1910 X	7.83	=	9.33
5	$(1 + 0.06)^4$ =	1.2625 X	7.83	=	9.89
6	$(1 + 0.06)^5$ =	1.3382 X	7.83	=	10.48
7	$(1 + 0.06)^6$ =	1.4185 X	7.83	=	11.11
8	$(1 + 0.06)^7$ =	1.5036 X	7.83	=	11.78
9	$(1 + 0.06)^8$ =	1.5938 X	7.83	=	12.48
			Total	=	90.00

hours worked in a year divided by the total machine hours which the machine is expected to deliver forms a depreciation charge factor. Multiplying this factor by the depreciation amount $I - L$ gives the depreciation charge during a particular year. With this method, depreciation charges vary with the production volume, providing a more stable tax rate for the firm but giving the accounting department a more difficult task.

Comparison of Depreciation Methods

Different depreciation methods provide different schedules of depreciation charges and book values. These differences may be illustrated by reviewing Tables 3-1, 3-3, 3-6, 3-7, 3-8, and 3-9 which show how various depreciation methods affect the accounting depreciation process with the same example. Changes in the book value for this example are summarized graphically in *Figure 3-2*.

Additional Comments About Depreciation Methods

The fixed percentage or declining balance method may be used for tangible property having a life of three years or more. It is generally restricted to new assets only. What holds for the fixed percentage method regarding new assets also holds for the sum-of-years'-digits method.

In addition to the quicker recovery on investments allowed since 1954, the Small Business Tax Act of 1958 was enacted to provide for an additional first year 20% depreciation allowance on new or used tangible personal property acquired after December 31, 1957. The property involved must have a useful life

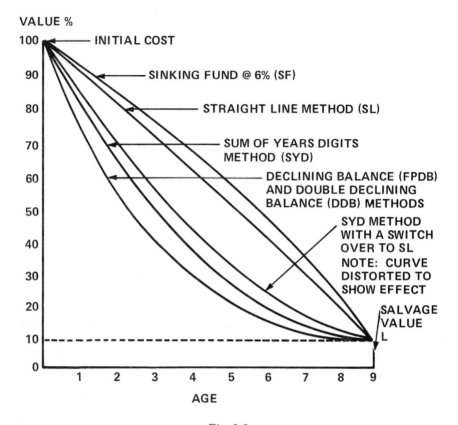

Fig. 3-3

of not less than six years and on a cost of not more than $10,000. The 20% allowance on the first year is in addition to the regular depreciation on the balance. A 1962 law allowed a 7% tax credit on new machinery in addition to depreciation. The investment tax credit laws change frequently, but the above are typical.

Calculating depreciation allowance is a difficult problem. First, an asset's life must be estimated, then its salvage value, if any, must be estimated. The Internal Revenue Service published Bulletin "F" in 1931 to guide business in determining average estimated life of a great many items of depreciable assets. Although this Bulletin has been superseded, it is still helpful. The problem of determining estimated life is accentuated by obsolescence resulting from the introduction of new and more economic machines for rendering the same service. Obsolescence can be defined as the total of all disadvantages (other than in operating cost) which exist in using an old asset which would not be present if a new one were available.

The salvage value of an asset 15 years hence must be estimated at the time the asset is purchased. With all the variables involved, *i.e.*, inflation, deflation, changes in design, changes in labor rates, etc., the problem of arriving at a realistic salvage value becomes almost insurmountable.

When depreciation allowances are inadequate for any period of time, several serious results can occur. Depreciation expense is charged into cost of sales and becomes part of the cost of a manufactured product. Therefore, if the depreciation charge is too low or inadequate to properly recover the capital invested, then income is overstated and taxes are overpaid. Thus, if taxes are overpaid due to overstated income, the overpayment comes out of capital and not out of earnings. This sad state of financial affairs is generally not recognized until replacement of assets becomes necessary. Then it is discovered that provision for replacement is inadequate and additional funds must be provided from some source other than the one reserved for depreciation (*i.e.*, replacement reserves).

Effect of Depreciation Methods on Profit-based Taxes

There are two significant items that indicate why the depreciation method affects taxes. They are:

1. Depreciation charges become part of replacement reserves

2. Replacement reserves are a source of internally generated capital, in addition to retained earnings, which are reinvested in interest-generating projects.

Consequently, the higher the depreciation, the higher the replacement reserves and the more funds available for internal investment. However, if more depreciation is taken early in the life of equipment, then less depreciation is taken later so that the extra depreciation funds which were deferred for internal use must eventually be decreased by an equal deficit. Thus, the choice of a fast writeoff depreciation method means only the use of extra funds for a period of time, but during this period of time, interest can be earned on the funds and this interest is retained, at least the portion not paid to tax.

Let us assume, for the sake of illustration, that the average corporate rate-of-return is 15% and that the firm is in the 40% tax bracket. Let us further assume that the firm is considering different depreciation methods and is using the preceding example to evaluate depreciation methods. An economic analyst might choose the SL depreciation method, the currently used method, as a basis of comparison to find the difference in depreciation changes between the SL method and each of the other methods. These algebraic differences can then be converted to an equivalent annual worth—or cost when the sign is negative. (See Tables 3-1, 3-3, 3-6, 3-7, 3-8, and 3-9.) The resulting equivalent annual value is the average before-taxes value which the firm realizes by switching to the depreciation method in question. Of course, the after-taxes annual return is 100%–40% or 60% of the before-taxes annual return. These calculations are given in Table 3-10 and their results show the realized gain per year by simply using depreciation methods with a faster early writeoff. Thus, a faster writeoff postpones the payment of some tax funds and the firm can obtain the interest which these funds can command over the period of postponement.

ECONOMICS OF MACHINE TOOL PROCUREMENT

Table 3-10. Time Series for FPDB

t	$(SPPW)_t^{15}$	X	$(CR)_9^{15}$	=	Total
1	0.8696	X	0.20957	=	0.182242
2	0.7561	X	0.20957	=	0.158456
3	0.6575	X	0.20957	=	0.137792
4	0.5718	X	0.20957	=	0.119832
5	0.4972	X	0.20957	=	0.104198
6	0.4323	X	0.20957	=	0.090597
7	0.3759	X	0.20957	=	0.078777
8	0.3269	X	0.20957	=	0.068508
9	0.2843	X	0.20957	=	0.059581

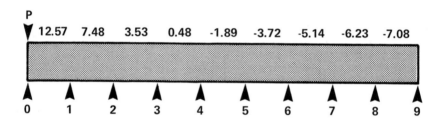

Fig. 3-4

Note in Table 3-10 that when $12.57 is discounted to the present worth amount at point P in the time series, the calculation is:

P = F (Single payment present worth factor for one year)

= 12.57 X 0.8696

= 10.930872

This amount is then time averaged for nine years by use of the uniform capital recovery factor as follows:

A = P (Uniform capital recovery factor)

= 10.930872 (0.20957)

= 2.290783

= 2.29

THE DEPRECIATION METHOD DECISION

This procedure is followed for the remaining years with the results as shown in Table 3-11.

Summary

Although depreciation has the connotation of value loss in common usage, this value depreciation concept is distinct from the concept of accounting depreciation which is a systematic process of allocating accounting charges for the purpose of recovering past cash outflows over time. Consequently, the market value of an asset and its accounting book value are rarely the same and this feature must be recognized in economic analyses such as replacement studies.

A number of terms that occur in depreciation accounting are defined, including: depletion, book value, sunk cost, depreciation charge, depreciation fund, depreciation service period, depreciation method, replacement reserves, and depreciation rate.

The purposes of depreciation accounting are record keeping of capital consumption and allocation, and capital recovery allocations. Such records aid in product pricing. They provide a basis for estimating the internal capital supply. And they provide audit information for financial control. Accounting depreciation stems from the need to replace equipment for physical or economic reasons.

A depreciation method is a rule which specifies the depreciation charge for each in the service life. One of the best known depreciation methods is the straight-line or SL method in which exactly $1/N$th of the depreciation amount is charged each year over the N years in the service period.

The fixed percentage of a declining balance depreciation method or FPDB consists of charging a constant percent r against the book value at the end of the previous year. However, in order for the FPDB method to provide a book value equal to the salvage value at the end of the service period, the value of r is strictly determined by the initial and salvage costs and the length of the service period.

A similar-behaving but simpler method for calculation than the FPDB method is one which is called the sum-of-the-years'-digits or, more simply, sum-of-the-digits or SYD method. A depreciation charge factor is developed in this method which is only dependent upon the length of the service period and provides the depreciation charge as a product with the depreciation amount $I-L$.

The double declining balance or DDB method is similar to the FPDB method except in this method the percentage r' is made to equal exactly twice the SL depreciation charge when multiplied by the initial value I. The results of using r' instead of the r percentage found in the FPDB method necessitates a switch-over to the SL method for the final year in order to bring the book value down to the salvage value at the service period's end. Other depreciation methods may also involve a switch-over to the SL method during the last half of the service period. The sinking fund or SF method differs from other depreciation methods in its use of interest in building the depreciation fund over the service period.

Table 3-11. Annual Savings of Example by Switching from the Straight-line

Difference between SL method charges and those of other methods.

Method	FPDB	SYD	DDB	Switch SYD	SF
Table	3-3	3-6	3-7	3-8	3-9
Year					
1	12.57	8.00	12.22*	8.00	−2.17
2	7.48	6.00	7.28	6.00	−1.70
3	3.53	4.00	3.44	4.00	−1.20
4	0.48	2.00	0.46	2.00	−0.67
5	−1.89	0.00	−1.87	0.00	−0.11
6	−3.72	−2.00	−3.67	−5.00	+0.48
7	−5.14	−4.00	−5.08	−5.00	1.11
8	−7.23	−6.00	−6.17	−5.00	1.78
9	−7.08	−8.00	−6.61	−5.00	2.48

Average annual gain over SL depreciation

*The legal limit is $20.00 − $10.00 = $10.00 for the first year rather than $22.00 − $10.00 = $12.22. This would change the charge difference from $2.23 to $1.82; the average gain to $1.76; and the 60% of annual gain to $1.06.

THE DEPRECIATION METHOD DECISION

Depreciation Method to Other Depreciation Methods*

Single payment present worth factor for year indicated times. the capital recovery factor for nine years	Time adjusted annual value of depreciation charge difference in the year at $i = 15\%$				
$(SPPW)_n^i$ $(UCR)_9^i$ $i - 15\%$	FPDB	SYD	DDB	SWITCH SYD	SF
0.182242	+2.29	+1.46	+2.23*	+1.46	−0.40
0.158456	+1.19	+0.95	+1.15	+0.95	−0.27
0.137792	+0.49	+0.55	+0.47	+0.55	−0.17
0.119832	+0.06	+0.24	+0.06	+0.24	−0.08
0.104198	−0.20	0.00	−0.20	0.00	−0.01
0.090597	−0.34	−0.18	−0.33	−0.45	+0.04
0.078777	−0.40	−0.32	−0.40	−0.39	+0.09
0.068508	−0.42	−0.41	−0.42	−0.34	+0.12
0.059581	−0.42	−0.48	−0.39	−0.30	+0.15
	+2.25	+1.81	+2.17*	+1.72	−0.53
	FPDB	SYD	DDB	Switch SYD	SF
= 60% of Annual gain =	+1.35	+1.09	+1.30*	+1.03	−0.32

Although the SF method calls for the present value of equal annual costs over the service period, the depreciation charges reflect the interest accumulation from the present to the time in which they are charged. Some other depreciation methods base their depreciation charges as a proportion of the actual or approximated output of the equipment to the estimated total output expected over the service period. These depreciation methods differ in the pattern of depreciation charges levied over the service period; and therefore they differ in the book value ascribed to the equipment, some methods resembling the typical value depreciation of equipment. In addition to the fact that some accounting depreciation methods provide better approximations to realizable market value, these methods differ in their effect upon profit-based taxes even though there is no change in the total amount of tax paid. The tax effect is strictly in the use of tax funds over a period of time and the realization of the interest which the use of this tax money generates. Therefore, the depreciation methods with faster writeoffs have a tax advantage over other depreciation methods.

CHAPTER 4

DECISIONS OF CAPITAL BUDGETING

Companies retain funds for purposes of reinvestment in new projects that require the acquisition of new assets. These assets are an investment which raises the company's capital value. Increased earnings are required through a rate of return from this added investment. Funds which are intended for reinvestment are **programmed expenditures**. It is a management function to decide *which* projects to elect out of many alternatives and *when* to provide the capital outlay for each elected project.

These capital budgeting decisions often represent the most crucial area of decision-making within a company because it is through these decisions that the company is able to move into new products, services, and processes. Further, they enable the company to take advantage of advancing technology, and to provide new facilities and equipment as additions or as replacements.

The amount of funds available for capital outlays depends on a company's ability to generate funds internally (*i.e.*, retained earnings) and to acquire funds externally (*i.e.*, borrowings, issuance of stock, and the like). Generally, a company will acquire enough funds to finance a number of different projects. The cost of acquiring this new capital is generally an amalgam of all the interest rates that must be paid to its lenders, both long and short run, and to its stockholders, both present and future.

In order to produce cash and develop capital budgets, a sales prediction must be made along with estimated costs for labor, materials, and overhead expenses. Estimates of income—which incorporate some judgements on the company's ability to generate funds—can then be made. The amount of internally generated funds can be estimated from financial statements and trends in sales and costs. Many companies would supplement their internal capital funds through borrowing, provided the return on the use of such funds would be greater than the cost. These companies may also increase their funds by retaining a higher percentage of their earnings; the problem of determining the percentage of earnings to retain is sometimes called the dividend problem.* Regardless of the methods used to generate money for capital outlays, the decision of how "best" to spend this money is of critical importance.

Cutoff Rate

The amount of money that a company can generate internally is usually insufficient to meet internal demands for capital. However, the rate of return generated by internal funds tends to be uniform. If prospective projects are elected in order of the estimated rate of return they generate, then the limited

---

*See D. W. Miller and M. K. Starr, *Executive Decision and Operations Research*, Prentice Hall, 1960, pp. 328ff (see Chapter 14 first), and see M. J. Gordon and E. Shapiro, "Capital Equipment Analysis; The Required Rate of Profit," *Management Science*, Vol. 3, No. 1, October 1956.

internal funds cut off the elected projects so that the least rate of return of the elected projects or the internal rate of return (whichever is higher) forms the cutoff rate of return as illustrated in *Figure* 4-1.

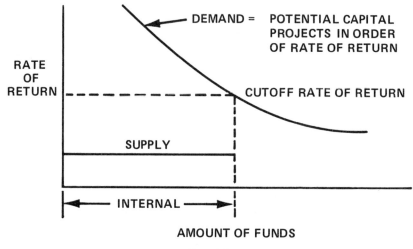

Fig. 4-1

The supply of funds can be extended when external sources of funds are considered with the cost which must be paid for their use, but with a rising rate of return required due to the added financing costs, *Figure* 4-2.

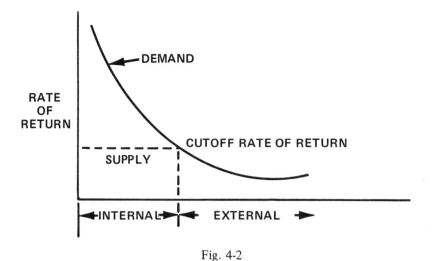

Fig. 4-2

When both internal and external financing are considered, the cutoff rate in capital budgeting can be defined as *the rate where the estimated supply of capital is just equal to the estimated demand for capital.* However, the cutoff rate beyond which projects can be excluded can be approached in several ways.

CAPITAL BUDGETING, METHOD 1. One method can be described as follows: The estimated rate of return is calculated on all the capital projects submitted, usually by the discounted cash flow method. The cutoff point is determined by the total budgeted capital outlay and the highest rates of return.

For example, let us assume that a company is able to generate $280,000 for capital expenditures from both internal and external sources. Let us further assume that a list of projects for capital outlays is submitted, including their prospective rates of return. From Table 4-1, the projects with the highest rate of return and with a cumulative total of $280,000 would be chosen. In their proper order, projects E, D, and B would be selected.

Table 4-1. Project Selection According to Rate of Return

Project Designation	Rate of Return (%)	Capital Outlay ($)	Cumulative Total Capital Outlay ($)
A	13.6%	100,000	100,000
B	24.0%	50,000	150,000
C	17.3%	100,000	250,000
D	31.5%	200,000	450,000
E	38.4%	30,000	480,000
F	9.7%	80,000	560,000

In practice, the sum of the project expenditures seldom equals the total capital budgeted. So either the total budgeted figure or the individual amounts in the projects must be changed. This may not be a formidable obstacle, but it should not be overlooked. It may also be necessary to accept many capital projects that do not conform to the expected rate of return. These projects would include capital expenditures required through new government regulations, requirements for safety, recent flood damage, new parking regulations for company personnel, and the like.

CAPITAL BUDGETING, METHOD 2. A second method of capital budgeting is as follows: A minimum acceptable rate of return is established; all projects which meet this standard are approved, provided sufficient capital funds are available. If 15% is set as the minimum acceptable rate of return, then projects B, C, D, and E in the preceding example would be approved for consideration.

In practice, considerations other than the maximum rates of return must often be recognized. Risk, governmental regulations, length of investment period, storm damage, and water damage are additional factors that may preclude a maximum rate of return decision.

Length of the investment period is well worth the time spent for a thorough analysis. For example, assume that the minimum acceptable rate of return is established at 25%. One project develops a rate of return of 27% for five years. A second project develops a rate of return of 23% for ten years. Depending on whether future projects offer more attractive opportunities, it may be better to stay with short-lived assets. Conversely, if future opportunities are somewhat depressed, it may be better to invest in a project giving less than the standard rate of return and in one with longer life.

Generally, companies develop a minimum rate of return that is sufficient to justify the risks and uncertainties of time and business conditions. Some examples will help to clarify this. The examples are calculated for rates of return before taxes. This does not mean that the influence of taxes is unimportant. With the present tax rate on industry, the tax effect on the rate of return is about 50%. This means that a 20% rate of return before taxes would be reduced to approximately 10% after taxes—and this is a very substantial reduction. However, the introduction of taxes does not alter the method of calculation, although it may alter the capital expenditure decision.

Let us consider several prospective projects with their expected cash flows. Negative flows indicate investments by the company and positive flows show profits over current costs as returns. With this information available, an analyst can calculate the rate of return. For example, the calculations of Project A are itemized in Table 4-2.

This form of calculation may be provided for every prospective project and summarized as indicated in Table 4-3.

It is common practice among many companies to establish flexibility in rates of return for the purpose of screening proposed projects. The apparent reason for flexibility in the rate of return is the inclusion of a risk factor which the rate of return reflects in order to make a project acceptable. In Table 4-3, for example, Project C does not generate positive flow until the sixth year. Thus, Project C may be highly risky, especially in a time of rapidly changing technology. Also, Project E with its high rate of return (38.4%) may not be as acceptable as Project D with 31.5%. If management feels that investment opportunities four years hence will return less than 30.0%, it may prefer to invest in the longer-lived project. But Project D ties up $200,000 of capital for fifteeen years, while Project E ties up capital of $30,000 for only four years. Using the payback period,* however, capital funds would be returned for Project D in approximately three and a half years and in Project E in two years.

Thus, many facets in the ever changing investment field must be investigated before sound budget decisions can be made. A good sound capital expenditure today may not be as good or as sound two years hence. Business conditions change with time, so it would be folly to make capital decisions on a purely quantitative basis. The quantitative analysis provides guidelines, but many factors such as the temper of the marketplace, the opportunities for investment, the general economic outlook, and others may completely alter the profile of

---

*See Chapter "Rate of Return Principles" for more information on the payback period.

capital expenditures. These reasons suggest that flexibility of thought in capital decision-making should be the key word.

Table 4-2. Calculations for Project A

End of Year	Cash Flow	gFfP* Factor 12%	15%	Present Worth 12%	15%
0	−100,000				
1	0	0.8929	0.8696	0	0
2	+ 5,000	0.7972	0.7561	3,986	3,781
3	+ 10,000	0.7118	0.6575	7,118	6,575
4	+ 30,000	0.6355	0.5718	19,065	17,154
5	+ 30,000	0.5674	0.4972	17,022	14,916
6	+ 30,000	0.5066	0.4323	15,198	12,969
7	+ 30,000	0.4523	0.3759	13,569	11,277
8	+ 30,000	0.4039	0.3269	12,117	9,807
9	+ 30,000	0.3606	0.2843	10,818	8,529
10	+ 30,000	0.3220	0.2472	9,660	7,416
				108,553	94,424

*Give future sum, find present sum

By linear interpolation between 12% and 15%,

$$\text{Rate of return} = 12\% + \frac{\$ 8,553}{\$16,629} (15\% - 12\%)$$
$$= 13.6\%$$

Analysis of Cash Flow Differences
When decisions involve the relative estimated profitability between two proposals, it is wise to recognize the relevancy of differences between alternative projects. An analysis of these differences between projects greatly simplifies the necessary calculations.

For example, assume that the cash flows of two projects are estimated and that the objective is to find the one most profitable for a given business enterprise. As in the previous examples, outflows will be designated as (−) and inflows as (+). Further assume a minimum rate of return of 20% before taxes and a five-year life or as it is sometimes called, a five-year **study period** of each project. (The relevant data are itemized in Table 4-4.)

Table 4-3. Summary of Cash Flow Examples

End of Year	\multicolumn{6}{c}{Projects}					
	A	B	C	D	E	F
0	−100,000	−50,000	−100,000	−200,000	−30,000	−80,000
1	0	+30,000		+ 50,000	+10,000	+ 5,000
2	+ 5,000	+20,000		+ 60,000	+20,000	+10,000
3	+ 10,000	+10,000		+ 70,000	+20,000	+10,000
4	+ 30,000	+10,000		+ 70,000	+20,000	+20,000
5	+ 30,000	+10,000		+ 70,000		+20,000
6	+ 30,000		+100,000	+ 70,000		+30,000
7	+ 30,000		+100,000	+ 70,000		+30,000
8	+ 30,000		+100,000	+ 70,000		
9	+ 30,000			+ 80,000		
10	+ 30,000			+ 80,000		
11				+ 80,000		
12				+ 80,000		
13				+ 80,000		
14				+ 80,000		
15				+ 80,000		
Rate of Return	13.6%	24.0%	17.3%	31.5%	38.4%	9.7%

It can easily be seen, that in Table 4-4, at 0% the cash flow of the differences is negative and equal to −$55,000. Interpreted, this means that an advantage of $55,000 accrues to Project A. Of course, if $55,000 were positive, then Project B would appear to be more attractive. However, the 20% minimum attractive rate of return applied to the differences enhances the attractiveness of Project A, as will be shown.

The difference between −$73,566 at 20% and −$55,000 at 0% reflects the different values of future dollars when compared with present dollars through the discounting process. At 0% interest, all future dollars are equal to present dollars.

The same results could be obtained by finding the present worth of cash flows for each of the two projects, as shown in Table 4-5. As these calculations show, the present worth of Project A is −$139,132 and the present worth of Project B is −$212, 698. This means that the present worth of Project A has a $73,566 advantage over Project B.

Table 4-4. Cash Flows for Two Projects

Year	Project A	Project B	B-A	20% P.W. Factor	Differences Present Worth
0	−$200,000	−$300,000	−$100,000	1.0000	−$100,000
1	+ 10,000	+ 20,000	+ 10,000	0.8333	8,333
2	+ 15,000	+ 20,000	+ 5,000	0.6944	3,472
3	+ 20,000	+ 30,000	+ 10,000	0.5787	5,787
4	+ 30,000	+ 40,000	+ 10,000	0.4823	4,823
5	+ 40,000	+ 50,000	+ 10,000	0.4019	4,019
					−$ 73,566

Table 4-5. Another Version of Cash Flows for Two Projects

Year	A Cash Flow	Factor	Present Worth	B Cash Flow	Factor	Present Worth
0	−200,000	1.0000	−200,000	−300,000	1.0000	−300,000
1	+ 10,000	0.8333	8,333	+ 20,000	0.8333	16,666
2	+ 15,000	0.6944	10,416	+ 20,000	0.6944	13,888
3	+ 20,000	0.5787	11,574	+ 30,000	0.5787	17,361
4	+ 30,000	0.4823	14,469	+ 40,000	0.4823	19,292
5	+ 40,000	0.4019	16,076	+500,000	0.4019	20,095
Totals			−139,132			−212,698

The Study Period

In the preceding examples it is assumed that both projects have equal service lives. When the project lives are different, some simplifying assumptions are often adopted in order to compare the alternatives. One of the simplest assumptions is to estimate that any replacement asset will have exactly the same cost pattern as the initial asset. A decision must then be made regarding the length of the study period on which comparisons will be made. As an example, it can be assumed that Project A was estimated to have a service life of four years, and Project B was estimated to have a service life of five years. What should the length of the study period be for which present worths will be compared? In cases of this type, it is sometimes convenient to extend the study period to a least common multiple of time, such as twenty years for the preceding example. Infinite study periods may also be employed through the use of continuous compound interest.

Effect of Taxes on Capital Budgeting

Taxes are a cost which a company pays to a government (city, county, state, or federal) for the privilege of doing business in the government domain as well as for the direct and indirect benefits received from government. However, the role of taxes as a cost means that the profit to the company—and the subsequent dividends which the company is capable of paying—are affected by the amount of tax cost. In addition, the relative profitability associated with various investment projects depends upon the tax basis and exemptions. There is a vast array of tax rates, bases, and exemptions in the modern business environment. Further, the array is changeable and variable. Accordingly, it can be demonstrated that the interpretation of the tax laws requires expert consultation, since taxes affect the capital budgeting process.

A common type of tax levied against a firm is a fixed or variable percentage of a company's residual revenues after the cost of sales and allowable expenses are removed. These residual revenues are more commonly called the net profit, which is the taxable income basis for taxes such as the federal corporate income tax. Variable tax rates are commonly achieved with this type of tax by providing a normal tax rate of x percent on taxable income and a surtax rate of y percent on taxable income in excess of some fixed quantity of taxable income. For example, let us assume a normal tax rate of 22% and a surtax rate of 26% applied to taxable income in excess of $25,000. In this example, the total tax paid is:

If the net profits exceed $25,000,
Tax paid = 0.22 (taxable income) + 0.26 (taxable income − $25,000)
Tax paid = 0.48 (taxable income) − $6,500
Otherwise,
Tax paid = 0.22 (taxable income)

Examination of this type of taxation reveals that the combined tax rate depends upon the net profits of the company. More profitable companies pay as high as 48% and low-profit firms as low as 22%. These tax rates change frequently.

A sliding-rate tax of the type just discussed aids in maintaining dividend stability throughout fluctuations in sales, but it creates greater computational complexity than the fixed-rate tax in assessing the tax obligations.

Fixed rate profit-based taxes clearly imply to those involved in the capital budget process that the realized rate of return must be inflated by the tax rate in the before-tax budgeting calculations. Thus, if a company is currently paying a 48% tax rate and it needs to realize at least a 10% rate of return in order to maintain an adequate capital supply, then the minimum project cutoff rate must be calculated as follows:

Cutoff rate of return (1 − tax rate) = Realized rate of return

$$\text{Cutoff rate} = \frac{0.10}{1-0.48} = \frac{0.10}{0.52} = 0.1925 = 19\% \text{ (approx.)}$$

Similar determinations are needed for the variable tax rate problem as long as the accepted budget projects do not affect the anticipated net profits. If, however, there is an anticipated increase in net profits associated with the election of an additional project, the realizable profits from the added project must bear the burden of the total tax increase. This increase is called a **project tax rate**. A project tax rate is the total increase in tax paid as a result of adding the project (*i.e.*, tax difference with and without project) divided by the taxable income from the project or:

$$\text{Project tax rate} = \frac{\text{Tax paid with project} - \text{Tax without project}}{\text{Taxable income from project}}$$

This project tax rate may be used in lieu of the tax rate for the purpose of finding the inflation:

$$\text{Project minimum rate of return} = \frac{\text{Realized rate of return}}{1 - \text{project tax rate}}$$

If the firm, for example, is currently netting a profit of $81,250, it is paying an income tax of:

Tax paid = 0.48 ($81,250) − $6,500 = $39,000 − $6,500 = $32,500

which amounts to an effective tax rate of:

$$\text{Tax rate} = \frac{32,500}{81,250} = 0.40 = 40\%$$

A project which is expected to increase the net income by $10,000 requires an additional tax of:

Tax paid = 0.48 ($91,250) − $6,500 = $43,800 − $6,500 = $37,300

and the new project tax rate is:

$$\text{Project tax rate} = \frac{\$37,300 - \$32,500}{\$10,000} = \frac{\$4,800}{\$10,000} = 0.48 = 48\%$$

Since this hypothetical company needs to earn a 12% rate of return after taxes, this new project must realize a before-tax rate of return of:

$$\text{Minimum rate of return} = \frac{0.12}{1-0.48} = 23\%$$

Another aspect of a project which has a decided tax effect is the determination of the **net profit**. Net profit is calculated as,

Net profit = Gross revenue − (cost of sales + allowed expenses)

Consequently, the greater the costs in the project which can be legally allocated to cost of sales or allowed expenses, the lower the net profit. Allowable expenses of this type include all or part of depreciation, and interest on borrowed funds. Sometimes applied and basic research expenses are included as a way to encourage research by reducing its attendant risks. Therefore, projects with a higher proportion of allowable expense components may be accepted at a lower before-tax rate of return and still provide the same after-taxes rate of return.

Two other taxes should be considered briefly in this discussion because of their effects on various projects. These taxes are based on **real** and **business property**. Consequently, projects which require heavy investments in land, buildings, equipment or inventories must be expected to offer a higher rate of return on the investment than projects which do not. Rate of return handicaps due to these forms of tax vary radically with the appropriate tax laws.

Summary

Management control of a company is in a large part effected through the programming of investments to projects within the company or to capital budgeting. The amount of capital to be invested depends upon the company's ability to generate internal capital; it also depends on the availability and cost of outside funds. The investments in the capital budget must be able to return the investment with sufficient interest to meet the demands of ownership and the need for company growth. Consequently, a rate-of-return screening device is used in capital budgeting in order to help select the more profitable projects. If a fixed supply of capital is available, the project rates of return form an ordered list for project selection. In cases where outside financing is available, the company may elect to take on additional projects. Sometimes flexible rates of return are employed to incorporate project risks. Cash flow differences may also be used as a basis of analysis for electing projects; however, problems in establishing common study periods are frequently encountered with this form of analysis.

The effect of taxation and the differential burden imposed upon prospective projects by various types of taxes are significant problems associated with project appraisal. Profit-based taxes require an inflated cutoff rate of return in budgeting in order to obtain a specified after-taxes rate of return. If the tax rate has a sliding scale, projects which extend net profits must bear the burden of the full extra tax rate. Other bases for taxation affect the tax costs of proposed projects differently, and these differences should be reflected in the expected returns from the projects.

CHAPTER 5

RATE OF RETURN PRINCIPLES

Many formulas embodying various interpretations of rates of return have been used in the engineering and financial fields for a great many years. Some of these formulas are still in use, yet many of them are based on straight average data. Most of these older, simpler methods do not take into consideration the economic value of time. Some of the more recent methods recognize the time value of money.

Some Rate of Return Methods
The first method to be considered is the rate of **return on first cost**.

$$\text{Rate of return} = \frac{\text{Average yearly profit}}{\text{First cost}}$$

This method isolates the profit attributable to the asset in question and averages this profit over the life of the asset. Neither the time value of money nor the capital wastage over the life of the asset are considered. This method could be used with some degree of reliability where different investment possibilities have the same life pattern and the cost patterns are relatively the same.

The second method is **return on average cost**.

$$\text{Rate of return} = \frac{\text{Average yearly profit}}{\text{Average first cost}}$$

This method is similar to the first cost method in that the numerator is the average yearly profit earned over the life of the asset. The denominator, however, is the sum of the undepreciated balance at the end of each year divided by the number of years of the asset's estimated life. This method takes into account the capital wastage, but does not consider the time value of the money amounts.

The third method relates to the payout period.

If the *number of years required to pay off the first cost* is represented by Y,

$$Y = \frac{\text{First cost}}{\text{Average profit/year}}$$

or

$$Y = \frac{\text{First cost}}{\text{Average profit/year} + \text{Average depreciation/year}}$$

or

$$Y = \frac{\text{First cost}}{\text{Average profit/year} + \text{Average depreciation/year} + \text{Average interest/year}}$$

These methods develop the number of years it will take the asset to pay off its first cost.

In addition to average profit, the denominator may have a provision for depreciation or depreciation and interest. The primary purpose of this is to allow for asset recovery of wastage and for interest return on the investment in the asset. These methods, however, do not recognize the economic value of time. This is an important disadvantage.

The fourth method relates to **present worth**. This method recognizes the time value of money. In its simplest form it consists of a time series represented by the first cost of the asset at zero time and the net difference each year between receipts and expenditures throughout the asset's life. The net difference, yearly, between the receipts and expenditures is discounted at some standard interest rate. The interest rate that equalizes the discounted difference of receipts and expenditures at zero time and the first cost of the asset at zero time is the so-called **true rate of return on the investment**.

The difference between the present worth of revenues and the first cost at zero time becomes a dollar profit or loss on the investment. A profit exists if the present worth of revenues is greater than the first cost. Conversely, an apparent loss exists if the present worth of revenues is less than the first cost at zero time. This, of course, assumes a particular standard interest rate that the investor wishes to use, *i.e.*, the company's expected capital rate or some rate acceptable to the company.

Sometimes a ratio of the present worth of the sum for all revenues to the first cost is used as an index of profitability. When the ratio of the present worth of revenues to first cost is unity at some specified interest rate, the investment earns exactly the interest rate used. When the ratio is greater than unity, the investment earns more than the interest rate used. Conversely, when the ratio is less than unity, the investment earns less than the interest rate used. The present worth method is similar to the rate of return method (which will be described next) in that a ratio of unity indicates the true rate of return on the investment.

This method recognizes the time value of money, since the revenues earned during the life of the asset are discounted to zero time at some standard interest rate.

The fifth and last method is the **interest rate of return**. This method provides the maximum rate at which the first cost is just equal to the sum of the present worths of all the revenues at zero time. Therefore, to calculate the rate of return for an asset, a trial-and-error approach is used. For example, a 10% rate might be applied to find the sum of the present worth of revenues. The present-worth sum may be greater than the first cost of the asset. If it is, then a 12% rate might be applied to find the sum of the present worth of revenues. This sum may be somewhat smaller than the first cost at zero time. A straight-line proportionality between 10 and 12% will then approximate the true rate of return between 10 and 12% without appreciable error.

RATE OF RETURN PRINCIPLES

Examples
RETURN ON FIRST COST. An asset has a useful life of 5 years, a first cost of $1000 and an average yearly profit of $200. Straight-line depreciation is used with a zero salvage value at the end of 5 years. The income tax rate is 50%. Cost of money is 10%.

$$\text{Rate of return} = \frac{\text{Average yearly profit}}{\text{First cost}}$$

$$= \frac{\$200}{\$1000}$$

$$= 20\%$$

RETURN ON AVERAGE FIRST COST. The average book value and average yearly profit is needed for this method. The depreciation on a straight-line basis is $1000 ÷ 5 = $200/year. The average book value is $500/year.

$$\text{Rate of return} = \frac{\text{Average yearly profit}}{\text{Average first cost}}$$

$$= \frac{\$200}{\$500}$$

$$= 40\%$$

Both averages are taken over the life span of the asset. However, when the tax life and the useful life of the asset are different, the preceding average method causes variations in the rate of return. In addition, where depreciation policies vary, the return is highest for the asset with the fastest writeoff.

PAY OUT PERIOD. This method is the reciprocal of the first two methods, and, in effect, is not a rate of return method. The actual return occurs when the first cost has been recovered. Some managements use this method as a measure of risk, indicating a number, such as three years, to determine whether to make the investment.

Again, letting Y represent *years to pay off first cost*,

(a) $$Y = \frac{\text{First cost}}{\text{Average profit/year}}$$

$$= \frac{1000}{200} = 5 \text{ years}$$

(b) $$Y = \frac{\text{First cost}}{\text{Average profit year} + \text{Average depreciation/year}}$$

$$= \frac{1000}{200 + 200} = 2.5 \text{ years}$$

(c) $Y = \dfrac{\text{First cost}}{\text{Average profit/year + Average depreciation/year + Average interest/year}}$

$= \dfrac{1000}{200 + 200 + 60} = 2.18$ years

Table 5-1. A Sample Calculation of Average Interest

Years	First Cost	Depreciation	Interest at 10%
1	$1000	$200	$100
2	800	200	80
3	600	200	60
4	400	200	40
5	200	200	20
			5/$300

$(P - L)(\frac{i}{2})(\frac{N+1}{N}) = 1000 = (\frac{10}{2})(\frac{6}{5}) = \$60$

The average interest is calculated as shown in Table 5-1. The average interest in this example is at the median point, *i.e.*, the third year.

This method is used in a number of different ways, and each one gives a different pay-back period. If a company regards three years as its established standard of risk, then Method 3a would reject the investment while Methods 3b and 3c would indicate acceptability. Generally, management has difficulty in arriving at a realistic pay-back period. After some pay-back standard is determined, the standard itself often prevents management from investing in profitable projects.

PRESENT WORTH METHOD. This method can best be illustrated by use of a time series.

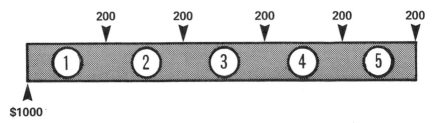

Fig. 5-1

RATE OF RETURN PRINCIPLES

Present worth factors will have to be used to bring the annual profits of $200 per year to zero time. The present worth of this series is computed at the standard interest rate indicated of 10%, as shown in Table 5-2.

Table 5-2. Sample Calculation of Present Worth

Year	Present Worth Factor 10%	Annuity Constant (Yearly Profit)	Present Worth Amount
1	0.9091	$200	$181.82
2	0.8264	200	165.28
3	0.7513	200	150.26
4	0.6830	200	136.60
5	0.6209	200	124.18
		Present Worth Sum	$758.14

The present worth sum of the yearly profits is $758.14. This compares with a first cost of $1000. Therefore, based on a standard interest rate of 10% as set by management, this project will lose approximately $242 ($1000 − $758).

INTEREST RATE OF RETURN (TRIAL AND ERROR). This method is also known as the discounted cash flow and profitability index method. Methods of this type sometime use continuous discount factors to arrive at present worth amounts. It differs from the present-worth method in the handling of the interest rate. The particular interest rate (rate of return) is calculated from the data of the problem as shown in Table 5-3.

A straight-line interpolation is made between 15% and 20%. Since the sum of the present worths is between the first cost of $1000, this calculation is as follows:

$$0.05 \left(\frac{1006 - 1000}{1006 - 897}\right) = \frac{6}{109} \times 0.05$$

$$= \frac{0.30}{109} = 0.0028$$

$$= 0.28\%$$

The 15% is added to the 0.28% for a total of 15.28% rate of return.

The examples used for illustration in this chapter are very simple. The real difficulty lies in estimating the revenues, the tax effect, the realizable values, obsolescence, and other real effects throughout the life of an asset. Further, the effect on profits, as they appear on the accountants' books, do not resemble the calculations used in the rate of return method. The profit and loss statements are

Table 5-3. Calculation of Rate of Return

Year	Gross Profit (Before Depreciation)	Depreciation	Net Profit (After Depreciation)
1	$400	$200	$200
2	400	200	200
3	400	200	200
4	400	200	200
5	400	200	200

Year	Net Profit (After Depreciation)	Net Profit (After 50% Tax)	Cash Flowback (Depreciation)	Total Cash Flowback at "0" Interest
1	$200	$100	$200	$300
2	200	100	200	300
3	200	100	200	300
4	200	100	200	300
5	200	100	200	300

Year	Flowback at 0%		at 15%				at 20%		
1	$300	x	0.8696	=	260.88	x	0.8333	=	249.99
2	300	x	0.7561	=	226.83	x	0.6944	=	208.32
3	300	x	0.6575	=	197.25	x	0.5787	=	173.61
4	300	x	0.5718	=	171.54	x	0.4823	=	144.69
5	300	x	0.4972	=	149.16	x	0.4019	=	120.57
					1005.66				897.18

not sufficiently detailed to show the annual profitability of each investment. Instead, the total profits on all investments are shown. These investments belong to various age groups and various lines, so that it becomes an insurmountable task to keep records of expenses and revenues in the detail that is necessary.

It would seem, then, that the methods described in this chapter *should not be confused with historical profit and loss statements.* Instead, they should be used as a decision-making tool by management. These methods can help in the very difficult problem area of replacement of capital equipment, expansion of old buildings, investment in new buildings, investment priorities, measurement of risk, and the like.

An attempt to compare the results of each method is completely valueless because the bases used for the calculations are different. This is particularly true of the rate of return method—the only method where an attempt is made to incorporate annually the effect of taxes on the study throughout the life of the project. Also, it is unrealisitic to assume that yearly profits will be constant. If estimates of expenses and revenues throughout the life of an asset can be made with some degree of accuracy, the interest rate of return method would seem to produce a more realistic answer.

Revenues and expenses that vary year by year can be accounted for properly through the use of discount factors. This means that a proper recognition of current and future dollar amounts is made. For the same reason, proper evaluation of varying depreciation amounts can be made in the study.

An example problem can be used to illustrate and emphasize the preceding comments.

Example:
A manufacturing operation makes 450 dies annually. From these 450, about 70%, or 315, are left and right hand dies made for opposite sides of a product. The die studied is processed on two boring mills and a planer. An economic study was proposed to determine whether a numerical control machine tool should be purchased to replace the three conventional machines currently used. The summarized data are shown in Table 5-4.

Table 5-4. Cost Comparison—NC v. Conventional Machine Tools

Year	Item	Conventional	NC	Marginal (Loss) or Savings
1	Basic Costs	$350,000	$733,935	$(383,935)
2	Training Costs	--	3,456	(3,456)
3	Die Machining Cost	162,027	58,248	103,799
4	Set-up Time	18,800	9,400	9,400
5	Tooling Cost	140,000	70,000	70,000
6	Maintenance Cost	4,200	4,200	--
7	Depreciation Cost (Average—9 years)	38,888	78,130	(39,242)
			Yearly Savings*	$ 143,242

*Total yearly savings is sum of items 3, 4, 5, 6, and 7.

Assuming that the estimated savings of $143,957 per year would remain constant over the estimated economic life of the NC machine, then by a trial and error procedure, the following present worth equation can be solved for the rate of return i.

$$733,935 + \$3456 = (gAfP)^i_9 \ (\$143,957)$$

This equation can be shown by the following time series.

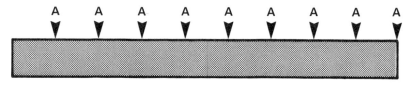

P = $737,391

Fig. 5-2

Where $A = \$143{,}957$ and the investment $P = \$737{,}391$

By substituting appropriate uniform present worth factors for 9 years, the i in the equation can be closely approximated. For example, the uniform present worth factor for 13% is 5.13. Then (5.13)($143,957) = $738,397. Simple linear interpolation can be used to get a return slightly greater than 13%. (In this study, the tax effect is not considered.)

Although many of the costs in this study were estimated by knowledgeable people, the economic justification for the purchase of an NC machine may be strengthened by emphasis of the following points:
1. The NC machine appears to have greater accuracy than the conventional machines it replaced
2. With NC, there is little need to worry about operator fatigue causing a costly error
3. Processing time has been dramatically reduced by the use of NC. Processing time reductions are very important in reducing leadtime.

Thus, it would seem profitable to buy the NC machine. The rate of return based on conventional machinery would approximate 13%.

CHAPTER 6

CASH FLOW MANAGEMENT

The term **cash flow** is probably the least understood of all management yardsticks. As the late Jules I. Bogen of New York University once noted, the term is hopelessly inaccurate: *Cash flow is neither cash nor flow.* The balance sheet of a rather small company can be of value in clarifying the term. In the following example, additions to liability accounts increase the cash effect and additions to asset accounts decrease the cash effect. Further, decreases to liability accounts decrease the cash effect and decreases to asset accounts increase the cash effect. This is illustrated in *Figure* 6-1, where it is seen that a decrease in cash balance was caused principally by the increase in accounts receivable from $183,174 to $287,927. The result was a cash flow decrease of $104,753.

For the purposes of this study, the machinery and equipment account, plus its contra account—the allowance for depreciation—should be analyzed. Apparently an equipment purchase of $42,152 was made during the period, resulting in a decrease in cash flow, but the depreciation account increases in cash flow by $12,257. Likewise, the depreciation expense account increased by this same amount, which flowed into cost of sales. This would have the effect of decreased gross profit and a decreased tax payable which would be shown in the income sheet. Part of this change would eventually appear in the earned surplus account in the balance sheet.

It should be noted that the *change* column should add algebraically to the $99,219. This amount must cross check with the difference bewteen $256,077 and $156,858—on line 54—cash on hand.

Some executives feel that the more they spend the more depreciation they create—and the more cash flow they generate. The final test, however, is whether capital expenditures generate more earnings and greater return on investment.

One must understand cash flow extremely well before making major decisions about capital investments. It would be well to increase this knowledge by a thorough study of financial ratios.

Financial Ratios
This is an audit technique used by management to determine how well a company is responding to the financial control imposed on it. It also aids in pinpointing problems or providing indications of needed corrections for capital budget management. The company's sales and earnings over extended periods of time may be compared to the gross national product (GNP) and of the industry average as a growth indicator. In order to more precisely locate the source of problems, companies typically investigate financial ratios, their trends, and how these ratios compare to those of other companies in the same industry. Thus, financial ratio analysis provides feedback information to management for use in capital budgeting and other aspects of financial control.

Figure 6-1. An Analysis of Cash Flow.

		BALANCE SHEET ACCOUNTS	START FISCAL YR.	CURRENT MONTH	CHANGE FISCAL YR.
Asset	1	US Treasury Bonds—Cost	150,978	150,000	+ 978
Asset	2	Accounts Receivable	183,174	287,927	−104,753
Liability	3	Allow. for Bad Debts	12,439	16,082	+ 3,643
	4				
	5				
	6				
Asset	7	Materials and Supplies	84,519	76,443	+ 8,076
Asset	8	Work in Process	33,928	27,636	+ 6,292
Asset	9	Finished Goods	51,604	00000	+ 51,604
	10				
	11				
Asset	12	Land	32,038	59,038	− 27,000
Asset	13	Bldg. and Improvements	92,437	92,437	000000
Liability	14	Allow. for Depreciation	40,981	43,314	+ 2,333
Asset	15	Mach. and Equip.	162,298	204,450	− 42,152
Liability	16	Allow. for Depreciation	84,974	97,231	+ 12,257
	17				
	18				
	19				
Asset	20	Notes and Advances	2,931	15,740	− 12,809
Asset	21	Prepaid Typewriter Exp.	11,857	1,081	+ 10,776
Asset	22	Unexpired Ins.		6,568	− 6,568
Asset	23	Deposits	4,040	2,702	+ 1,338
Asset	24	Deposits—Postage		238	− 238
Asset	25	Patents and Good-will	1	1	0
Asset	26	Pension Plan			
	27				
	28				
	29				

CASH FLOW MANAGEMENT

		BALANCE SHEET ACCOUNTS	START FISCAL YR.	CURRENT MONTH	CHANGE FISCAL YR.
	30				
Liability	31	Accounts–Trade Payable	35,048	51,298	+ 16,250
Liability	32	Customer Credits–Est.	74,424	61,860	– 12,554
Liability	33	Federal Inc. Tax Year	126,701	00000	–126,701
Liability	34	Royalties	91,220	9,827	– 81,393
	35				
	36				
	37				
Liability	38	Dividends	2,732	2,432	– 300
Liability	39	Payroll	14,188	4,804	– 9,384
Liability	40	Taxes	3,880	4,249	+ 369
Liability	41	Withholding Tax	4,805	3,752	– 1,053
Liability	42	Defense Bonds	84	340	+ 256
Liability	43	Insurance, Employees		910	+ 910
	44				
	45				
Liability	46	Federal Inc. Tax-est. Yr.		112,781	+112,781
	47				
Liability	48	Preferred Stock	198,700	191,100	– 7,600
Liability	49	Common Stock–$100 Par	271,496	271,496	000000
Liability	50	Earned Surplus	104,210	223,893	+119,683
Asset	51	Treasury Stock		14,250	– 14,250
	52				
	53				
	54	Cash on Hand	256,077	156,858	– 99,219

The operating concept of financial ratio analysis is simplicity itself; it consists of finding the company's ratio between costs in particular categories and recording this ratio with those obtained from the financial statements of other companies in the industry over time. Thus, financial ratio analysis provides a graphic representation of a company relative to its industry while also providing a basis for interpreting trends in the industry and divergent trends by the company.

A closely associated aspect of financial ratios is the identification of company growth over time in comparison to national growth. This analysis may be performed by graphing the gross national product over time along with the company's and its industry's average sales and earnings. A straight-line fit of each curve provides a simple basis for contrasting growth in the slope of the fitted lines. For example, if a 5.7% per year slope is found for the GNP, then the company's sales and earnings must also be 5.7% annually to keep pace with the national economy. If the industry's growth in sales and earnings is less than the GNP's rate, along with those of the company, then the diagnostic tool indicates that the firm should consider expansion into new fields of endeavor—fields which provide good growth. Should the industry, but not the company, exhibit good growth compared to the GNP, it can be concluded that the company has serious internal problems which an internal financial ratio analysis may help to disclose. Divergent sales and earnings trends provide indication of low profit product lines and other sources of difficulties which warrant internal investigation.

Financial ratio analysis for internal auditing may consider various types of cost ratios which are useful to management. These ratios are included in the list shown in Table 6-1. The ratios in this table are grouped according to the general class of characteristics to which the financial ratio applies. However, interpretations of the ratios' implications to management control require a more detailed description. These ratios are typical of the data provided by such financial firms as Dun and Bradstreet, so that industry averages, as well as industry high and lows, may be obtained and referenced as a comparison of a company's financial position to that of others in the field.

Table 6-1. Common Financial Ratios.

A. **Profit Potential**
   1. Sales/dollars of capital funds
   2. Sales/dollars of common equity
   3. Net profit/sales
   4. Sales/working capital
   5. Net profit/capital funds
   6. Net profit/net worth

B. **Operating Results**
   7. Operating profit/capital funds
   8. Operating profit/sales

9. Net sales inventory
10. Inventory/working capital
11. Fixed assets/net worth
12. Sales/fixed assets
13. Sales/employee

C. **Credit Standing**
14. Current assets/current liabilities
15. Net quick assets
16. Total debt/net worth
17. Funded debt/net working capital
18. Current liabilities/net worth
19. Common equity/capital funds

D. **Growth Record**
20. Sales/average sales
21. Net capital/average capital
22. R & D funds/net sales

E. **Stability**
23. Net profit/operating profit
24. % decline in earnings in lowest year/average earnings

F. **Payout Ratio**
25. Earnings/share
26. Earnings paid to common stock/earnings to preferred

G. **Per Share of Stock**
27. Sales
28. Dividends
29. Net asset equity
30. Net current asset equity

H. **Market Price**
31. Price/share
32. Sales/dollar of common stock
33. Earnings/dollar of common stock
34. Dividend yield
35. Net assets/stock price on market

Most of the ratios in Table 6-1 have obvious implications. A few will be cited to show how trends in some of these ratios indicate the need for corrective action.

Profitability is an important characteristic and yardstick. This characteristic may be viewed with respect to the marketing strategies employed through the **net profit to sales ratio** (#3) or it may be viewed with respect to project investments through the **net profits to net worth ratio** (#6); the latter is generally above 10%. The failure of these two ratios to keep a similar trend (slope) over time is indicative of a need to revise marketing or investment strategies.

Of particular importance to the production management personnel are the **sales to inventory** (#9) and **inventory to working capital** (#10) ratios. The first ratio indicates the firm's ability to correctly anticipate sales and maintain sufficient stocks to prevent prolonged stock surpluses or outages (*i.e.*, inventory balance). However, a balance must be maintained between the working capital tied up in unsold inventory and that required for current obligations. Consequently, the changes in ratio #10 can indicate a dangerous situation and ratio #9 can pinpoint the reason.

Some of the other ratios and their implications are discussed individually, as follows:

FIXED ASSETS TO NET WORTH (Ratio #11). Normally, this ratio should not exceed 100% for manufacturing concerns. Above 100%, the company has so much of its capital frozen into machinery or "bricks and mortar" that the necessary margin of operating funds for carrying receivables, inventories, and day-to-day cash outlays as well as maturing obligations becomes too narrow. This exposes the business to the hazards of unexpected developments such as sudden change in business climate. It may also create drains on income in the form of heavy carrying and maintenance charges should a serious portion of fixed assets lay idle for a substantial length of time.

CURRENT ASSETS TO CURRENT LIABILITIES (Ratio #14). This ratio is commonly known as the **current ratio** or the **banker's ratio**. It measures a company's ability to cover its current obligations with its current assets. A historically time-honored relationship of 2:1 is widely accepted. In other words, the current assets could be reduced 50% and still be sufficient to cover obligations to current creditors.

TOTAL DEBT TO NET WORTH (Ratio #16). A danger signal for this ratio may be 100%. When the ratio exceeds 100%, the equity of creditors in assets of the business exceeds that of the owners. The management of top-heavy liabilities entails strains and hazards which can become a threat to business survival. Judgment can become clouded, and anxieties which sap management's energies are often created. Additionally, the business is exposed to the risks of contingencies such as sudden downturns in sales, changes in style or customer preference, rapid rises in business costs, strikes, fires, and floods. These and a host of other factors can upset the operation of a modern concern that is handicapped with an unfavorable debt-to-net worth ratio. A concern which is carrying a large short-term or long-term debt under such circumstances generally faces rough going.

CURRENT LIABILITIES TO NET WORTH (Ratio #18). Often it appears as though companies suffer from lack of capital when their problem is actually the misuse or unsound use of capital. These companies are usually guilty of overtrading, *i.e.*, handling a volume of business too large for their net working capital. From an analyst's point of view, this excessive use of credit is nothing more than heavy liabilities, generally current. The measure of that use is the relationship between current assets/current liabilities and the current liabilities/ net worth. Companies which have heavy current liabilities need study and attention. Generally speaking, current liabilities should not exceed 3/4 of net worth.

Summary

Management control of a company is in a large part effected through the programming of investments to projects within the company or through capital budgeting. Of course, the amount of capital to be invested depends upon the company's ability to generate internal capital and on the availability and cost of outside funds. However, the investments in the capital budget must be able to return the investment with sufficient interest to meet the demands of ownership and the need for company growth. Consequently, a rate of return screening device is used in capital budgeting in order to help select the more profitable projects. If a fixed supply of capital is available, the project rates of return form an ordered list for project selection. In cases where outside financing is available, the firm may elect to take on additional projects down toward the cost of outside capital. (See *Fig.* 4-2.) Sometimes flexible rates of return are employed to incorporate project risks. **Cash flow differences** may also be used as a basis of analysis for electing projects; frequently, however, there are problems with this type of analysis in establishing common study periods.

A significant problem associated with the appraising of projects is the effect of taxation and the differential burden imposed upon prospective projects by various types of taxes. Profit-based taxes require an inflated cutoff rate of return in budgeting in order to obtain a specified after-taxes rate of return. If the tax rate has a sliding scale, projects which extend net profits must bear the burden of the full extra tax rate. Other bases for taxation affect the tax costs of proposed projects differentially and these differences should be reflected in the expected returns from the projects.

An audit technique which assists management in appraising how well the firm responds to the financial control imposed on it and aids further in pin-pointing problems or providing indications of the need corrections for capital budget management is a **growth and financial ratio analysis**. The firm's sales and earnings over time may be compared to that of the gross national product and of the industry average as a growth indicator. In order to more precisely locate the source of problems, firms typically investigate financial ratios, their trends, and how these ratios compare to those of other firms in the same industry. Thus, financial ratio analysis provides feedback information to management for use in capital budgeting and other aspects of financial control.

REPLACEMENT ALGORITHMS

Several methods are currently employed in industry to measure the profitability of an investment. In this book, the **discounted cash flow method** is suggested. This method calculates the rate of return by evaluating the time value of money. It recognizes interest as the cost incurred in the use of money, and therefore is judged to be the most realistic method to use for decisions regarding long-term investments, such as computers or numerically controlled machinery.

Because of the many variations among accounting systems and the wide range of computer and numerical control applications, no single procedure fits all investment situations. However, the discounted cash flow method to be described is adaptable to virtually any application. In particular, the guidelines and procedures for quantifying the specific impact of a proposed investment on individual operational expenses can be effectively employed in any traditional equipment-investment analysis method. As an example, every investment evaluation process requires answers to the following questions—answers which can be quantitatively developed by the procedures recommended in the following methodology:

1. What are the *individual elements* in the total investment?
2. What is the *amount required* for each investment factor?
3. *When* will the expenditure of funds be required?
4. What *direct operational savings* are anticipated from the investment?
5. What *indirect* operational savings are anticipated from the investment?
6. *How much* will be saved in each operational savings factor and *when* will the savings be realized?

The discounted cash flow method focuses on the amounts and timing of cash receipts and disbursements. Other methods might fail to give proper emphasis to the timing of cash flows, thereby incorrectly ranking competing investment alternatives. The significance of the timing of cash flows is illustrated in the following example.

The competing investments require the same initial disbursement of $300,000. Each has an expected life of ten years and zero salvage value. Option *A* is expected to yield a total profit of $450,000 over ten years, while a total profit of $500,000 is projected for Option *B*.

A tabulation of the year-by-year cash flows of both investments is given in Table 6-2. By using the discounted cash flow method, the rate of return on Option *A* is computed to be 20 1/2%, compared to a rate of return of 6 1/8% on Option *B*.* This should not be surprising when one carefully examines the cash flow patterns of both investments, shown graphically in *Figure* 6-2. A dollar in hand today has a greater present worth than a future dollar, since it can be reinvested for the coming year at the prevailing interest rate. Consequently, the early yearly profits of Option *A* weigh more heavily than the late profits of Option *B*. The various phases and evaluation methods are as shown in *Figure* 6-2.

---

*See the illustration of rate of return computations, p. 68.

Table 6-2. Annual Cash Flow for Two Competing Projects.

	Option A		Option B	
Year	Annual Profit ($1000)	Cumulative Profit ($1000)	Annual Profit ($1000)	Cumulative Profit ($1000)
1	200	200	0	0
2	100	300	0	0
3	50	350	10	10
4	30	380	10	20
5	20	400	20	40
6	20	420	20	60
7	20	440	40	100
8	10	450	50	150
9	0	450	125	275
10	0	450	225	500
Total Profit	$450,000		$500,000	

PHASE I—*Determining the total cost of the proposed investment.* Because expenditures are required to implement certain operational and organizational modifications and because there is a need for new programming and maintenance skills and for supporting equipment, the total cost of an initial machine can be considerably more than just the equipment price. Some of these expenses need not be duplicated when additional equipment is installed, however.

PHASE II—*Estimating the savings.* Computer and numerical control operational savings are evaluated according to direct and indirect classifications. Direct savings accrue from the economies of part floor-to-floor production time, *i.e.*, the time it takes to complete jobs on a conventional machine minus the time required to produce the same jobs with computerized or numerically controlled operations. The method employed to estimate direct savings differs little from the techniques used for conventional equipment. Direct savings can be converted to dollars by multiplying the time saved by the appropriate labor and overhead rates for the setup and machine operator personnel.

Indirect operational savings occur in cost categories other than machine setup and operator expense. Some of these savings are in maintenance, tooling,

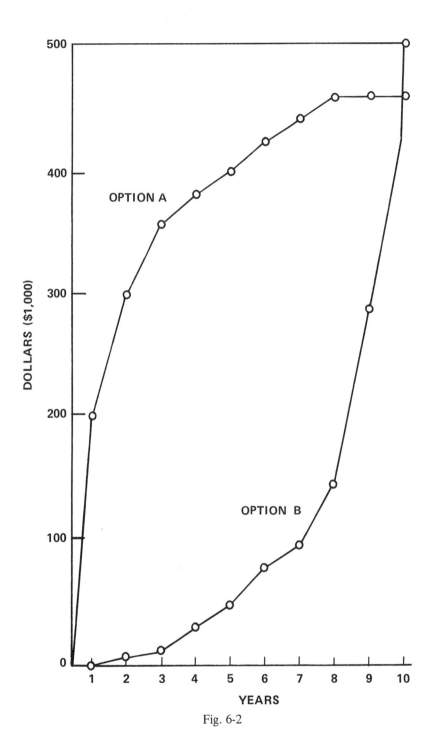
Fig. 6-2

programming, inspection, scrap, inventory, floor space, material handling, etc.

PHASE III—*Computing the discounted cash flows and determining the equivalent rate of return.* Future cash flows anticipated from the investment and the annual savings are discounted, where appropriate, and an interest rate representing the rate of return is computed. In investment applications, the resulting rate of return is the compound interest rate at which the predicted savings and revenues are equivalent to the required disbursements. It can also be described as the interest rate at which the total present value of a series of expected annual savings is equivalent to the initial investment.

The resulting rate is then compared with the minimum rate of return acceptable for a capital investment. The minimum rate determined by management is influenced by many factors, including the prevailing cost of capital and the company's willingness to assume financial risk.

A simplified example with hypothetical figures will help to illustrate how the discounted cash flow technique is used to compute the rate of return for computer or numerical control investments. The following conditions are assumed:

Investment

1. The equipment costs $50,000.
2. Training cost for maintenance, programming, and operating personnel is $2000; shipping and installation is $1000; and additional supporting equipment is $7000.
3. Therefore, the initial total investment outflow for Items 1 and 2 is $60,000.

Savings

1. Assume that numerical control makes possible direct operational savings of 1250 machine hours per year. Assuming an operator cost of $10 an hour, in a one-man, one-machine situation, the direct savings amount to $12,500 per year.
2. Assume that the indirect operational savings are:

Maintenance (negative savings)	$(400)
Tooling	6500
Programming (negative savings)	(2500)
Inspection	1000
Scrap	900
Inventory Carrying Costs	1000
Net Indirect Savings	$6500

3. The total annual direct and indirect savings are then $19,000 per year. If the tax rate is 50%, the after-tax savings are $9500 per year.
4. The net effect of the depreciation charge is an annual cash inflow equal to the depreciation charge multiplied by the tax rate. Assuming a ten-year machine life with a uniform rate of deterioration and a straight-line depreciation method, in this example the after-tax inflow would be $6000 times 0.50 or $3000. Adding this to the net annual savings of $9500 yields a net annual cash inflow of $12,500.

Computing the Rate of Return

1. The discounted cash flow technique determines the interest rate at which a cash inflow of $12,500 per year for ten years (the estimated economic life of the equipment) will be provided from the "annuity" investment of $60,000 today.
2. Compound interest tables are used to determine the interest rate that provides the ten annual payments from the initial investment of $60,000. In other words, the analyst determines the interest rate that provides ten yearly "annuity" payments of $12,500 from a present investment of $60,000. This interest rate is best calculated by trial and error. Two interest rates are assumed, then the present worths or equivalent annual payments are calculated, after which the interest rate can be found by interpolation.

Using this method, the present worth of the savings series of $12,500 for ten years is equal to $60,000 at the rate of approximately 15%. The rate of return in this example is therefore 15%.

The foregoing is an overview of one effective computer or numerical control investment evaluation process.

CHAPTER 7

THE GENERAL REPLACEMENT ALGORITHM

Engineering economists have published many papers and books about the **replacement algorithm**. Basically, this algorithm is designed to provide decision-makers with an opportunity to select the best investment from several alternatives. The result can be expressed as a before or after-tax rate of return.

Before any quantitative results can be obtained, however, certain operating, maintenance, and capital costs must be collected or estimated. This can be a difficult and sensitive task because many companies do not keep individual operating and maintenance costs on their various machine tools. In addition, it is difficult to obtain operating and maintenance costs on new machines that a company does not own, but may purchase. Some of the pertinent costs that must be analyzed are:
1. Initial investment cost
2. Shipping and installation costs
3. Foundation costs (if any)
4. Initial tooling costs
5. Direct labor costs
6. Maintenance costs
7. Inspection costs
8. Scrap costs
9. Floor space costs
10. Material costs.

When cost comparisons for the replacement algorithms are made, the cost differences become the important factors. For example, a drilling machine and a milling machine may be candidates for replacement by one numerically controlled machine tool. Direct labor costs may be reduced by 50% because the numerically controlled machine may require only one operator, while the drilling and milling machines may require two operators. In like manner, cost differences must be calculated for each item of cost that is analyzed and considered pertinent to the replacement analysis.

Typically, the minimum cost replacement interval is easy to construct graphically, but it is difficult to supply meaningful data to the model. This is especially true when new machines are involved in the decision. The graphic minimum replacement model is depicted in *Figure* 7-1.

General Replacement, Minimum Cost Model

The operating and maintenance cost curve is of some interest. For a number of tools studied, the break-in and debugging costs start at a relatively high cost level and then gradually decrease. This break-in period generally lasts for less than a year.

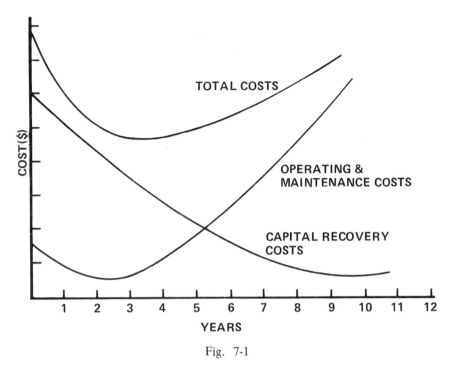

Fig. 7-1

The model shown in *Figure* 7-1 can be used to develop minimum costs for both the present and proposed equipment. To calculate minimum costs, however, it must be assumed that past costs of the present equipment can be collected and future costs of the proposed equipment can be predicted with a fair degree of accuracy.

To develop data for the model, it is desirable to use present-worth and capital-recovery factors. Doing so allows for recognition of the time value of money. In essence, it is simply an application of the principles of compound interest. The use of this economic discipline reflects two factors: the cost of funds and the risk.

The **cost of funds** is a variable. A dollar received today has a greater value than a dollar received one year from now. This discounting is accomplished through the use of **present-worth** factors. In addition, once the total present worth of a time series is known, time-adjusted annual costs can be calculated through the use of **capital-recovery** factors.

Risk may be defined in the present context as an exposure to loss. This loss should be evaluated in monetary terms, if possible. For instance, in a manufacturing process, historical data may reveal that one piece out of 132 pieces does not pass quality-control inspection. The money loss of the rejected piece was found, for example, to be $86. Thus, the probable loss due to rejection is $0.65 per piece. In this example, the risk is reasonably known and may be considered as an input cost.

When the risk is unknown, it may be estimated and added as an input cost or it may be considered an irreducible factor in making the decision.

Assuming that the probabilities of several possible futures can be estimated, the decision-maker faces the problem of buying, new equipment—or not buying new equipment. The uncertainty of estimated cost data is part of the risk. The uncertainty of estimating the useful economic life of the equipment is also part of the risk. Other risk factors include the investment of funds in this particular category, socio-economic problems, and technological change and obsolescence. Many other risk factors could be suggested; those mentioned are merely indicative.

The MAPI Replacement Algorithm

One of the standard methods for evaluation of machine tool purchases is the MAPI procedure. The letters MAPI refer to the Machinery and Allied Products Institute. This institute, under its research director George Willard Terborgh, published *Dynamic Equipment Policy* (McGraw-Hill, 1949), which described the application of the MAPI procedure. A later publication of this procedure was published in 1958 by Mr. Terborgh under the title *Business Investment Policy*. Both books are recommended reading, since space requirements do not permit adequate treatment of the MAPI procedure.

Briefly, the concept makes use of expressions such as operating inferiority, adverse minimum, and basic assumptions.

Operating inferiority develops on almost any currently used machine when it is compared with the best new identical machine available at any point in time. Mr. Terborgh uses the words **deterioration and obsolescence** rather than depreciation, because depreciation is commonly employed to cover wear and tear and obsolescence. Deterioration and obsolescence vary considerably between various types of machine tools. Some machine tools suffer heavy depreciation but little obsolescence. Whatever combination of these factors exists, an operating inferiority develops between the machine presently in service and the best available machine at the point in the time horizon under consideration.

The **adverse minimum** consists of operating inferiority and capital cost. The objective is to find the proper proportion between the two in order to minimize their sum.

The two assumptions in theory are:
1. Future challengers will have the same adverse minimum as present challengers.
2. The present challenger will develop operating inferiority at a constant rate over its estimated service life.

Thus, the decision to replace is based on comparisons of the adverse minima between challenger and defender. *Figure* 7-2 graphically illustrates the procedure.

The adverse minimum is represented approximately at Point A on the combined curve. By definition, it is the lowest point on this curve. It should be pointed out that when making actual evaluations by use of the MAPI algorithm and the MAPI work sheets, the first year's rate of return is used as the basis for making the "buy" decision. If a planning horizon of more than one year is required for the replacement decision, the MAPI procedure generally would not be used.

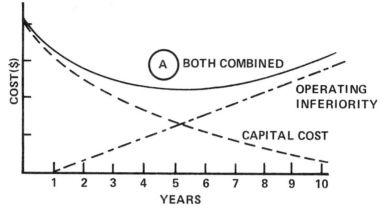

Fig. 7-2

A General Replacement Actual Example

In any comparative study between two or more alternative purchases, similar activities should be used (wherever possible) for the cost comparisons. For example, capital costs should be compared with capital costs, operating costs with operating costs, etc. Also, it is somewhat meaningless to compare one plant's costs with another's because various cost factors that are valid in one plant may be entirely different or nonexistent in another. Therefore, one should not use the cost experience of another plant to estimate one's savings. Experience indicates that economic investment studies should be made on a plant-to-plant basis.

Let us consider the example shown in *Figure* 7-3. This is a small aluminum piece requiring some drilling and counterboring. The plant making this part was ideal for study, since both conventional and numerically controlled machines were available. Only a small number of parts were produced per setup as would be typical in a job shop.

A small conventional drill press was compared with a small NC model.

To determine the relative cost to operate each machine, the following information was obtained:
1. Capital Costs
   a. Initial investment
   b. Cost of tooling and accessories
   c. Original setup cost
   d. Depreciation method
   e. Interest on investment (including risk)
   f. Expected life of machine
2. Maintenance Expense
3. Operating expense
4. Direct labor expense
5. Overhead expense
6. Incidental costs

THE GENERAL REPLACEMENT ALGORITHM

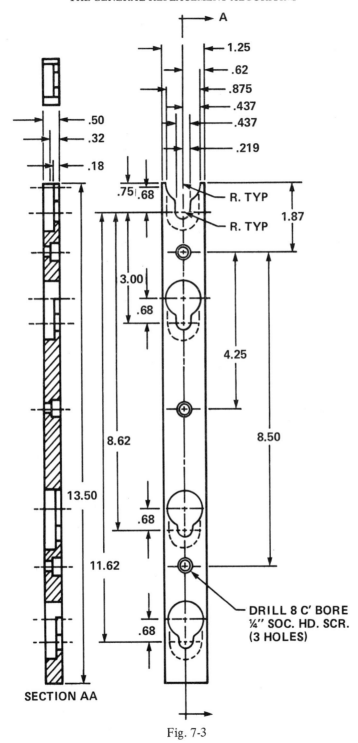

Fig. 7-3

Additionally, it was felt that the following information should be obtained:
1. Physical dimensions of part
2. Specifications of machining operations (tolerances, feeds, speeds, etc.)
3. Operating method of the man-machine system (motion pattern of the operator, workplace layout, etc.).

The cost data listed in Table 7-1 were obtained from the cost records of the company. The operating time data were obtained by stopwatch measurement, by predetermined time systems, and by time-study measurements previously made by the company. The company using these machines calculated their investment decision based on a *20 percent rate of return before taxes.*

Quantitative Procedure

The economic comparative cost analysis procedure is not complicated. Use is made of various compound interest formulas, which discount values in a time series to the same time base, usually the present. The six most common interest formulas in use today are as follows:

1. $F = P(1 + i)^n$ ... Single payment—future worth

2. $P = \dfrac{F}{(1 + i)^n}$ ... Single payment—present worth

3. $A = F\left[\dfrac{i}{(1 + i)^n - 1}\right]$ ... Annuity—sinking fund

4. $F = A\left[\dfrac{i}{(1 + i)^n - 1}\right]$ ... Annuity—compound amount

5. $A = P\left[\dfrac{i(1 + i)^n}{(1 + i)^n - 1}\right]$ ... Annuity—capital recovery

6. $P = A\left[\dfrac{(1 + i)^n - 1}{i(1 + i)^n}\right]$ ... Annuity—present worth

where P = present amount, F = future amount, A = annuity constant, i = nominal rate of interest, and n = time periods. In this study, Formulas 2, 5, and 6 will be used, but not necessarily in that order.

These formulas are used extensively in economic cost and replacement studies. Compound interest tables can be purchased in virtually any bookstore. For a reference book having a representative sample of compound interest tables, it is suggested that E. Grant and Ireson, *Principles of Engineering Economy*, 5th Edition, Ronald Press, be used.

The first calculation is to transform the capital costs of depreciation and interest on the investment to equivalent uniform annual costs for both the numerically controlled machine and the conventional machine. The calculations are presented in Tables 7-2 through 7-7.

THE GENERAL REPLACEMENT ALGORITHM

Table 7-1. Cost Data—Manual v. NC Machines

Manual Machine	Cost
a. Original Cost	$ 2,250
b. Freight in	85
c. Tooling	500
d. Backup tools	500
Total	$ 3,335
e. Maintenance	600/yr.
f. Direct labor	3.50/hr.
Fringe benefits	.35/hr.
Note: 43-hour workweek	
52 weeks/year	
g. Overhead allocated to machine	
Floor space @ 1.00/sq. ft.	150
Property tax	50
Power, heat, light, etc.	300
Total	$ 500/yr.
h. Expected life—7 years (no salvage value)	
i. Double-declining balance depreciation used.	
The percentage calculation follows:	

$$\text{Rate} = \frac{200\%}{7} = 28.57\%$$

Numerically Controlled Machine	
a. Original cost	$20,493
b. Freight in	120
c. Tooling	495
d. 5-horsepower air compressor	495
e. Installation	200
Total	$21,803
f. Maintenance	1,800/yr.
g. Direct labor (same as manual)	3.85/hr.
h. Overhead allocated to machine	
Floor space @ 1.00/sq.ft.	400/yr.
Property tax	400/yr.
Power, heat, light, etc.	420/yr.
Total	$ 1,220/yr.
i. Expected life—7 years	
j. Double-declining balance depreciation used.	

Table 7-2. Total Estimated Time-adjusted Capital Costs (Conventional).

n	A Dollars Realizable at First of nth year	B Dollars Depreciation in nth year	C Interest (20%) on Realizable Value at First of nth Year($)	D Depreciation Plus Interest in nth Year ($)	E Present Worth Factor
1	3,335	953	667	1,620	0.8333
2	2,382	681	476	1,157	0.6944
3	1,701	486	340	826	0.5187
4	1,215	347	243	590	0.4823
5	868	248	174	422	0.4019
6	620	187	124	301	0.3349
7	443	127	89	216	0.2791

Table 7-3. Total Estimated Time-adjusted Capital Costs (NC).

n	A Dollars Realizable at First of nth year	B Dollars Depreciation in nth year	C Interest (20%) on Realizable Value at First of nth Year($)	D Depreciation Plus Interest in nth Year ($)	E Present Worth Factor
1	21,803	6,229	4,361	10,590	0.8333
2	15,574	4,449	3,115	7,564	0.6944
3	11,125	3,178	2,225	5,403	0.5787
4	7,947	2,270	1,589	3,859	0.4823
5	5,677	1,622	1,135	2,757	0.4019
6	4,055	1,159	811	1,970	0.3349
7	2,896	827	579	1,406	0.2791

THE GENERAL REPLACEMENT ALGORITHM

F	G	H	I	J
Present Worth of Capital in Costs nth Year ($)	Cumulative Present Worth for n Years ($)	Capital Recovery Factor	Time Adjusted Capital Costs Annual ($)	Time Adjusted Capital Costs ($) Hourly
1,350	1,350	1.2000	1,620	0.73
803	2,153	0.6545	1,409	0.63
478	2,631	0.4757	1,249	0.56
285	2,916	0.3863	1,126	0.50
170	3,086	0.3344	1,032	0.46
101	3,187	0.3007	958	0.43
60	3,247	0.2774	901	0.40

F	G	H	I	J
Present Worth of Capital in Costs nth Year	Cumulative Present Worth for n Years ($)	Capital Recovery Factor	Time Adjusted Capital Costs Annual ($)	Time Adjusted Capital Costs ($) Hourly
8,825	8,825	1.2000	10,590	4.74
5,252	14,077	0.6545	9,214	4.12
3,127	17,204	0.4747	8,167	3.65
1,861	19,065	0.3863	7,365	3.29
1,108	20,173	0.3344	6,745	3.02
660	20,833	0.3007	6,265	2.80
392	21,225	0.2774	5,888	2.63

Table 7-4. Total Estimated Time-adjusted Operating and Maintenance Costs (Conventional)

n	Operating & Maintenance Costs in nth Year ($)	Present Worth Factor (20%)	Present Worth in nth Years ($)	Cumulative Present Worth in nth Years ($)	Capital Recovery Factor (20%)	Time-Adjusted Costs ($) Operating & Maintenance	
						Annual	Hourly
1	9,700	0.8333	8,083	8,083	1.2000	9,700	4.34
2	9,900	0.6944	6,875	14,958	0.6945	9,791	4.38
3	10,100	0.5787	5,845	20,803	0.4747	9,876	4.42
4	10,300	0.4823	4,968	25,771	0.3863	9,955	4.45
5	10,500	0.4019	4,220	29,991	0.3344	10,028	4.48
6	10,700	0.3349	3,583	33,574	0.3007	10,096	4.52
7	10,900	0.2791	3,042	36,616	0.2774	10,158	4.54

THE GENERAL REPLACEMENT ALGORITHM

Table 7-5. Total Estimated Time-adjusted Operating and Maintenance Costs (Numerical Control)

n	Operating & Maintenance Costs in nth Year ($)	Present Worth Factor (20%)	Present Worth in nth Years ($)	Cumulative Present Worth in nth Years ($)	Capital Recovery Factor (20%)	Time-Adjusted Costs ($) Operating & Maintenance	
						Annual	Hourly
1	11,620	0.8333	9,686	9,683	1.2000	11,620	5.20
2	11,820	0.6944	8,208	17,891	0.6945	11,711	5.24
3	12,020	0.5787	6,956	24,847	0.4747	11,796	5.28
4	12,220	0.4823	5,894	30,741	0.3863	11,875	5.31
5	12,420	0.4019	4,992	35,733	0.3344	11,948	5.34
6	12,620	0.3349	4,226	39,959	0.3007	12,016	5.37
7	12,820	0.2791	3,578	43,537	0.2774	12,078	5.40

Table 7-6. Sum of Time-adjusted Capital Costs and Time-adjusted Operating and Maintenance Costs (Conventional)

n	Time Adjusted Operating & Maintenance Costs ($)		Time Adjusted Capital Costs ($)		Total Costs ($)	
	Annual	Hourly	Annual	Hourly	Annual	Hourly
1	9,700	4.34	1,620	0.72	11,320	5.06
2	9,791	4.38	1,409	0.63	11,200	5.01
3	9,876	4.42	1,249	0.56	11,125	4.98
4	9,955	4.46	1,126	0.50	11,081	4.96
5	10,028	4.48	1,032	0.47	11,060	4.95
6	10,096	4.52	958	0.42	11,054	4.94
7	10,158	4.54	901	0.40	11,059	4.94

Table 7-7. Sum of Time-adjusted Capital Costs and Time-adjusted Operating Costs (Numerical Control)

n	Time Adjusted Operating & Maintenance Costs ($)		Time Adjusted Capital Costs ($)		Total Costs ($)	
	Annual	Hourly	Annual	Hourly	Annual	Hourly
1	11,620	5.20	10,590	4.74	22,210	9.94
2	11,711	5.24	9,214	4.12	20,925	9.36
3	11,796	5.28	8,167	3.65	19,963	8.93
4	11,875	5.31	7,365	3.29	19,240	8.60
5	11,948	5.34	6,745	3.02	18,693	8.36
6	12,016	5.37	6,265	2.80	18,281	8.17
7	12,078	5.40	5,888	2.64	17,966	8.04

Table 7-8. List of Elements

No.	Element	Frequency	Element (Hours)	Conventional (Hours)	NC (Hours)
1.	Get job and drawing; study it.	1	0.0833	0.0833	0.0833
2.	Obtain necessary tools from crib.	1	0.0833	0.0833	0.0833
3.	Set up machine.				
	A. Clean table; wipe clamp surface.	1	0.0010	0.0010	0.0010
	B. Load and position vice (Moog).	1	0.0833	0.0833	0.0833
	C. Set depth stops (conventional).	1	0.0500	0.0500	
	D. Set depth stops (numerical control).	3	0.0500		0.1500
4.	Program and punch.	47	0.0125		0.5875
5.	Pick up piece; place in vice.	1	0.0008	0.0008	0.0008
6.	Tighten vice.	1	0.0006	0.0006	0.0006
7.	Tighten piece with mallet.	1	0.0005	0.0005	0.0005
8.	Tighten vice with mallet.	1	0.0005	0.0005	0.0005
9.	Place spotting drill in spindle (½ tool change).	1	0.0008	0.0008	0.0008
10.	Turn on coolant.	1	0.0005	0.0005	0.0005
11.	Start machine.	1	0.0004	0.0004	0.0004
12.	Position and spot centers.				
	A. Position-conventional.	11	0.0042	0.0462	
	B. Position-numerical control.	11	0.0008		0.0088
	C. Raise and lower spindle.	11	0.0005	0.0055	0.0055
	D. Brush chips aside.	2	0.0008	0.0016	0.0016
	E. Change stop depth.	11	0.0008	0.0088	0.0088
	F. Spot centers.	11	0.0008	0.0088	0.0088
13.	Stop machine.	1	0.0004	0.0004	0.0004
14.	Change tool to 9/32 drill.	1	0.0015	0.0015	0.0015
15.	Start machine.	1	0.0004	0.0004	0.0004
16.	Drill taps as centered.				
	A. Position-conventional.	11	0.0019	0.0209	
	B. Position-numerical control.	11	0.0003		0.0033
	C. Raise and lower spindle.	11	0.0005	0.0055	0.0055

ECONOMICS OF MACHINE TOOL PROCUREMENT

No.	Element	Frequency	Element (Hours)	Conventional (Hours)	NC (Hours)
	D. Drill	11	0.0002	0.0022	0.0022
	E. Brush chips aside.	11	0.0008	0.0088	0.0088
17.	Stop machine.	1	0.0004	0.0004	0.0004
18.	Change tool to 7/8 diam. end mill.	1	0.0015	0.0015	0.0015
19.	Start machine.	1	0.0004	0.0004	0.0004
20.	End-mill 0.875 diam. holes.				
	A. Position-conventional.	4	0.0019	0.0076	
	B. Position-numerical control.	4	0.0003		0.0012
	C. Raise and lower spindle.	4	0.0005	0.0020	0.0020
	D. Machine part.	4	0.0042	0.0168	0.0168
	E. Brush chips aside.	4	0.0008	0.0032	0.0032
21.	Stop machine.	1	0.0004	0.0004	0.0004
22.	Change tool to 7/16 end mill.	1	0.0015	0.0015	0.0015
23.	Start machine.	1	0.0004	0.0004	0.0004
24.	End-mill 7/16 hole and cut across to 7/8 diam. hole.				
	A. Position-conventional.	4	0.0019	0.0076	
	B. Position-numerical control.	4	0.0003		0.0012
	C. Raise and lower spindle.	4	0.0005	0.0020	0.0020
	D. Drill.	4	0.0024	0.0096	0.0096
	E. Cut (allow for 1" cut).	4	0.0017	0.0068	0.0068
	F. Brush chips aside.	4	0.0008	0.0032	0.0032
25.	Stop machine.	1	0.0004	0.0004	0.0004
26.	Set depth stop to 0.32".	1	0.0017	0.0017	0.0017
27.	Start machine.	1	0.0004	0.0004	0.0004
28.	End-mill 2 sec. hd. ser. holes to 0.32" depth.				
	A. Position-conventional.	3	0.0019	0.0057	
	B. Position-numerical control.	3	0.0003		0.0009
	C. Raise and lower spindle.	3	0.0005	0.0015	0.0015
	D. Machine part.	3	0.0016	0.0048	0.0048
	E. Brush chips aside.	3	0.0008	0.0024	0.0024

THE GENERAL REPLACEMENT ALGORITHM

No.	Element	Frequency	Element (Hours)	Conventional (Hours)	NC (Hours)
29.	Stop machine.	1	0.0004	0.0004	0.0004
30.	Change tool to 7/8 diam. end mill.	1	0.0015	0.0015	0.0015
31.	Release vice.	1	0.0006	0.0006	0.0006
32.	Turn piece over.	1	0.0004	0.0004	0.0004
33.	Tighten vice.	1	0.0006	0.0006	0.0006
34.	Tighten vice with mallet.	1	0.0005	0.0005	0.0005
35.	Tighten piece with mallet.	1	0.0005	0.0005	0.0005
36.	Start machine.	1	0.0004	0.0004	0.0004
37.	End-mill 7/8 diam. holes at 0.32" depth across opening.				
	A. Position-conventional.	4	0.0019	0.0076	
	B. Position-numerical control.	4	0.0003		0.0012
	C. Raise and lower spindle.	4	0.0005	0.0020	0.0020
	D. Cut part (allow 1" cut).	4	0.0017	0.0068	0.0068
	E. Brush chips aside.	4	0.0008	0.0032	0.0032
38.	Stop machine.	1	0.0004	0.0004	0.0004
39.	Release vice.	1	0.0006	0.0006	0.0006
40.	Remove piece and lay aside.	1	0.0007	0.0007	0.0007
41.	Remove 7/8 diam. end-mill (½ tool change).	1	0.0008	0.0008	0.0008
42.	Change depth stop.	1	0.0017	0.0017	0.0017
43.	Inspect piece by scale.				
	A. Conventional-every piece.	2	0.0013	0.0026	
	B. Numerical control-1 in 5.	2	0.0006		0.0012
44.	Count pieces.	1	0.0002	0.0002	0.0002
45.	Turn off coolant.	1	0.005	0.0005	0.0005
46.	Tear down machine (unclamp vice; clean table and tools; sweep area; return tools to crib).	1	0.2500	0.2500	0.2500

By summing the time-adjusted capital costs and the time-adjusted operating and maintenance costs, the total costs can be ascertained. One should note that the capital costs represent a decreasing time series, while the operating and maintenance costs represent an increasing time series. This is shown in Tables 7-6 and 7-7.

The dollars per running hour can be used to calculate the price cost when the standard time per piece is known. Standard time per piece can be determined by breaking the operation down into its elements and then time studying each element. These elements essentially determine the method for performing the operation. A complete list of the elements is assembled in Table 7-8.

With these element data, the length of time can be determined for producing any number of parts. The time elements can be divided into two categories: fixed elements and variable elements. The fixed elements are elements performed only once, regardless of the number of parts made; i.e., the same parts. The variable elements are elements performed once for every part produced. These element data can now be used to develop predicting equations for the operation—one for numerical control and one for conventional. The general predicting equation generally takes the following linear form:

$$T = F + VP$$

where T = Time (hours) to produce P parts, F = Fixed time (setup and teardown), V = Variable time/part, and P = Number of parts.

By synthesizing the basic element data, the time equations for the numerically controlled and conventional machines can be determined.

A. Fixed Time

	Time (hours)	
	Conventional	NC
1. Setup (Elements 1-4)	0.3009	0.9884
2. Teardown (Elements 45, 46)	0.2505	0.2505
Total Fixed Time	0.5514	1.2389

B. Variable Time

Operation (Elements 5-44)	0.2177	0.1346

C. Time Equations (see *Figure* 7-4)

$T_c = 0.5514 + 0.2177P$ —Conventional

$T_n = 1.2389 + 0.1346$ —Numerical Control

By graphing these equations, a comparison can be made of the productivity of each machine as a function of time. This graph is a typical break-even chart and is presented in *Figure* 7-4. The graph has an ordinate of *time in hours* and an abscissa of *number of parts produced*. Because the fixed time of the numerically controlled machine is relatively high, a batch must contain approximately eight pieces before the numerically controlled machine can break even with the conventional machine. It must be remembered that this statement is true only for this particular part when the described method of operation is used for producing it. It is conceivable that the batches of fairly large numbers, an entirely different tooling, fixturing, and workplace layout may be changed to reduce the operation time per piece. It should also be pointed out that the part in this study is relatively simple to machine.

For this particular operation, there were two major differences in method between the numerically controlled machine and the conventional machine. The

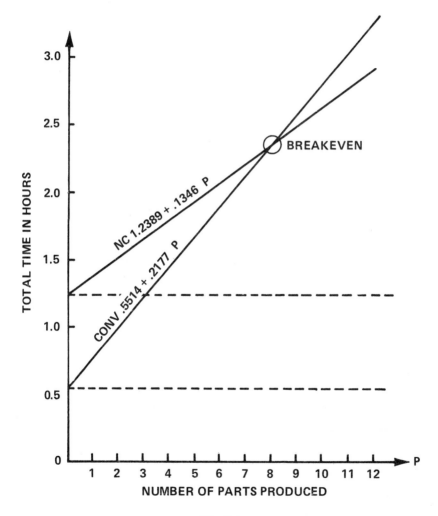

Fig. 7-4

numerically controlled machine had to be programmed and the positioning element time was much faster than the positioning element time for the conventional machine. Thus, one must consider the programming time addition as well as the faster positioning times in order to develop realistic answers to questions about possible savings between numerically controlled and conventional machines.

The Comparison
By using the operating, maintenance, and capital cost summary tables, the time-adjusted conventional cost is $4.94 per hour while the time-adjusted numerical control cost is $8.04 per hour. The total cost equations then become:

Conventional
$$C_c = 4.94\,(0.5514 + 0.2177P)$$
Numerical Control
$$C_n = 8.04\,(1.2389 + 0.1346P)$$

In these equations P represents the number of pieces.

A MAPI Actual Example
Suppose the problem is one of making decisions related to replacing inter-city truck tractors in the trucking industry so that operating costs are minimized through optimum timing of replacements. Stated another way, a method is to be developed to determine optimum service life applicable both to an individual vehicle and to an entire group of similar vehicles. The four major components of cost considered in this problem are capital cost, interest on investment, operating cost, and obsolescence.

The net capital outlay for an asset, over its lifetime, is the original cost, plus the cost of any original additions during the period of use, less the cash received (salvage value) for the asset when it is finally sold.

The effect of interest should be taken into consideration in replacement calculations. A replacement study is to a considerable degree a financial transaction since it is a study of the relative value of alternative series of cash payments and receipts. In computing capital cost, for example, if $10,500 cash is paid for a vehicle which is sold two years later for $4000, the net cash outlay has been $6500. The $10,500 has been tied up for two years, resulting in an additional loss of interest (perhaps at 15%) of $3386. This represents a payment, two years in the future, of $6500 plus $3386, or $9886. A payment of $9886 two years in the future has a present worth of $7477 at 15% and is also equivalent to an annual payment at the end of each year of $4598. Thus, $4598 is the true, or *time-adjusted* average capital cost for this transaction.

The only operating costs which need to be considered in a replacement study are those which change with the age (in miles or years) of the equipment. For tractors, these costs would include fuel and oil, regular repairs, and preventive maintenance.

For our purpose, obsolescence may be defined as the total of all disadvantages (except operating cost) which exist from using an old tractor and which would not be present if a new tractor were available. Regardless of the

THE GENERAL REPLACEMENT ALGORITHM

difficulty of placing a cash value on obsolescence, it must be reckoned as a fact of life.

When time-adjusted average costs per year for capital, operating inferiority, and obsolescence have been computed for various service lives at some selected interest rate, computation of the optimum service life then amounts to nothing more than selecting that service life in which the total average cost per year over the life of the asset is the smallest. Before we actually present the model example, a few comments about it are in order.

The firm replaced the tractors every nine years. The inital cost of each tractor was $10,500. The following are anticipated capital additions: a major overhaul costing $1400 (without interest) in maintenance labor, material, and overhead to be required after 200,000 miles; the installation of a new engine at a cost of $4000 (without interest) to be required after 400,000 miles; and another overhaul after 600,000 miles. The unit was assumed to be worthless on the open market after five years.

The operating costs considered were: fuel and oil, regular repairs, and preventative maintenance. The operating cost of a tractor over its service life less the operating cost which would be incurred if new equipment were available at all times is called the total operating inferiority of the old vehicle.

Obsolescence costs were based on the following factors:

1. Ability of a new tractor to pull heavier loads
2. Ability of a new tractor to make faster time on a given run
3. Ability of a new tractor to have fewer forced stoppages on the road.

Tables 7-9 through 7-12 show all the necessary calculations that provide the decision-maker with cost comparisons that help him to make better decisions. The first calculation is to transform the capital costs, operating inferiority, and obsolescence into equivalent (time-adjusted) uniform annual costs. These time-adjusted costs are added to get time-adjusted total costs. The calculations follow.

To select an interest rate to be used in replacement calculations it is generally necessary to take the following factors into consideration:

1. The average capital invested over the life of the asset as it would be shown on the company's books
2. The amount of debt acquired, if any, by the investment and the interest rate paid on the debt
3. The amount of funds supplied by the company for the investment
4. The after-tax rate of return desired by the company on its invested capital
5. The current income tax rate and the method of depreciation used for tax purposes.

For this problem an interest rate of 15% was assumed.

On the average, general changes in price level will affect all cost elements to approximately the same degree. Thus, over a period of time, general inflation may raise the average cost per mile without appreciably changing the optimum service life. Over a shorter time period, however, it will generally be found cheaper to delay replacement somewhat during a period of rising prices.

Table 7-9. Calculation of Time-adjusted Averages for Model Problem: Capital Cost

Year	(1) Salvage Value at End of Year	(2) Loss of Salvage During Year	(3) Interest On Opening Salvage Value (15%)	(4) Capital Additions Including Interest	(5) Total Capital Cost	(6) Present Worth Factor (15%)	(7) Present Worth	(8) Present Worth Accumulated	(9) Capital Recovery Factor (15%)	(10) Time-Adjusted Averages
1	$6,000	$4,500	$1,575	$ –0–	$6,075	0.870	$5,285	$ 5,285	1.150	$6,075
2	4,000	2,000	900	–0–	2,900	0.756	2,192	7,477	0.615	4,598
3	2,000	2,000	600	1,610	4,210	0.658	2,770	10,247	0.438	4,488
4	1,000	1,000	300	–0–	1,300	0.572	744	10,991	0.350	3,847
5	–0–	1,000	150	4,450	5,600	0.497	2,783	13,774	0.398	4,105
6	–0–	–0–	–0–	–0–	–0–	0.432	–0–	13,774	0.264	3,636
7	–0–	–0–	–0–	–0–	–0–	0.376	–0–	13,774	0.240	3,306
8	–0–	–0–	–0–	1,610	1,610	0.327	527	14,301	0.223	3,189
9	–0–	–0–	–0–	–0–	–0–	0.284	–0–	14,301	0.210	3,003

THE GENERAL REPLACEMENT ALGORITHM

Table 7-10. Calculation of Time-adjusted Averages for Model Problem: Operating Inferiority

Year	(1) Operating Inferiority	(2) Present Worth Factor (15%)	(3) Present Worth	(4) Present Worth Accumulated	(5) Capital Recovery Factor (15%)	(6) Time-Adjusted Averages
1	$ 0	0.870	$ 0	$ 0	1.150	$ 0
2	1,102	0.756	833	833	0.615	512
3	1,503	0.658	989	1,822	0.438	798
4	2,539	0.572	1,452	3,274	0.350	1,146
5	2,610	0.497	1,287	4,571	0.298	1,362
6	3,063	0.432	1,323	5,894	0.264	1,556
7	3,606	0.376	1,356	7,250	0.240	1,780
8	3,762	0.327	1,230	8,480	0.223	1,891
9	4,264	0.284	1,211	9,691	0.210	2,035

Table 7-11. Calculation of Time-adjusted averages for Model Problem: Obsolescence

Year	(1) Obsolescence	(2) Present Worth Factor (15%)	(3) Present Worth	(4) Present Worth Accumulated	(5) Capital Recovery Factor (15%)	(6) Time-Adjusted Averages
1	$ 0	0.870	$ 0	$ 0	1.150	$ 0
2	100	0.756	76	76	0.615	47
3	200	0.685	132	208	0.438	91
4	300	0.572	172	308	0.350	133
5	400	0.497	199	579	0.298	173
6	500	0.432	216	795	0.264	210
7	600	0.376	226	1,021	0.240	245
8	700	0.327	229	1,250	0.223	279
9	800	0.284	227	1,477	0.210	310

Table 7-12. Calculation of Time-adjusted Averages for Model Problem: Total Costs

Year	(1) Miles Per Year	(2) Time-Adjusted Capital Costs	(3) Time-Adjusted Operating Inferiority	(4) Time-Adjusted Obsolescence	(5) Total Time-Adjusted Costs	(6) Time-Adjusted Cost/Mile
1	100,000	$6,075	$ 0	$ 0	$6,075	6.08 ¢
2	100,000	4,598	512	47	5,157	5.16
3	90,000	4,488	798	91	5,377	5.57
4	90,000	3,847	1,146	133	5,126	5.40
5	80,000	4,105	1,362	173	5,640	6.15
6	80,000	3,636	1,556	210	5,402	6.00
7	70,000	3,306	1,740	245	5,261	6.45
8	70,000	3,189	1,891	279	5,359	6.40
9	60,000	3,003	2,035	310	5,348	6.50

Using a cost-per-mile data from the problem calculations (Table 7-12, Column 6) as a base, one finds that it will generally be profitable to keep vehicles for four years, perhaps even longer, if price increases are extremely steep. On the other hand, declining prices would probably make it more economical to replace every two years, providing the other assumptions in the model problem are reasonably accurate. The time-adjusted cost per mile for two years is 5.16 cents; for four years, 5.40 cents; and for nine years, 6.50 cents. Thus, by replacing the tractors every two years, assuming zero inflation, a savings of 1.34 cents per mile (6.50−5.16) can be achieved. (See Table 7-12.)

Summary of Additional Principles
Let us assume, for comparison's sake, that the machine decision example was given to five different people who each took a different approach to the problem and then met together to provide top management with a justified recommendation. Each man would give his solution to the group, estimating annual cost savings on the basis of whatever evaluation criteria he used as justification.

Of course, each decision-maker at the meeting would argue for his approach by attacking the weaknesses of other approaches. One should remember, however, that there is no universally accepted approach to making "best" decisions in the face of future uncertainty. Nevertheless, the principles described here provide some guidance in proceeding toward one's decision. The superior line of reasoning in uncertainty decisions is one in which the specific aspects of the decision are compatible with the decision-maker's attitudes toward gains, losses, and risk. This is difficult enough for an individual, but an individual decision-maker acting for an institution is faced with sharply increased possibilities of risks, losses, or gains. His responsibility will be a formidable, if not impossible, task. Clearly, balanced judgment is required.

The reader may also ponder if decisions arise very often with several states of nature clearly identified but with no ability to estimate each state's probability. This query has led the students of many of these principles to believe that uncertainty principles are an academic invention. Nevertheless, this discussion may pose several introspective questions regarding how the individual arrives at his decisions for a fallible future.

CHAPTER 8

OTHER INVESTMENT CRITERIA

Introduction

Many decisions are faced in engineering and management situations which have differing economic implications, depending upon some future state of affairs. If the state of affairs cannot be perfectly predicted at the time the decision is to be made, then the decision-maker must somehow resolve the decision with uncertainty. He must choose one of the available economical alternatives and recognize the fallibility of his decision. This discussion deals with decision-making problems in which a choice is made from a list of various courses of action. The decision-making process is described theoretically and the implications of using this theory in practical applications are noted. Decisions with uncertainty are common in the equipment replacement area.

A Formal Representation of a Decision

Decision-making is a process of selecting between various courses of action open to the decision-maker. These courses of action are called alternatives, and symbolized as a_i. The decision-maker considers a collection of n alternatives, a_1, $a_2 \ldots, a_i, \ldots a_n$, and he faces the task of selecting one of them. If the decision-maker is rational, he should select the alternative which is the best in *some* sense. However, the "best" alternative is dependent upon an evaluation scheme which attempts to predict the outcome of the various alternatives. The outcome, in turn, is dependent upon situations outside of the decision-maker's control. Such uncontrollable situations are often called **states of nature**, symbolized as s_j, and a collection of m mutually exclusive states are considered in the decision as $s_1, \ldots, s_2, \ldots, s_j, \ldots, s_m$. Together a_i and s_j are said to denote the outcome, Θ_{ij}, resulting from the decision-maker's selection of the ith alternative and nature's selection of the jth state. If some consistent evaluation scheme is applied to the outcome Θ_{ij} in the form U_{ij}, then outcomes can be compared so that the best alternative may be selected.

A compact representation of these decision components is provided by the format given in *Figure* 8-1. The decision-maker is assumed to look down the column of alternatives and select that one which he "believes" will result in the greatest value, the maximum U_{ij}, where:

a_i is any alternative course of action to be chosen by the decision-maker

s_j is a distinct state of nature, an uncontrollable condition such as the gradual wearing out of a machine tool or obsolescence caused by technological changes

Θ_{ij} is the outcome resulting from the decision-maker's choice of a_i and nature's choice of s_j

$U(\Theta_{ij})$ is the utility or value associated with the outcome Θ_{ij} in terms of costs or benefits.

ECONOMICS OF MACHINE TOOL PROCUREMENT

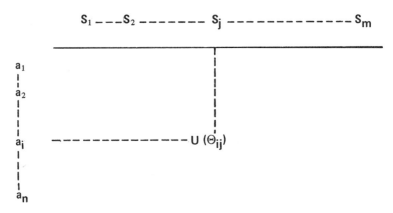

Fig. 8-1

A common example for illustrating the alternatives—the states-of-nature decision format—is the egg-omelet choice. The problem statement might be something like this:

> You have decided that you want a six-egg omelet and you have just broken five eggs and put them in a bowl. You are reaching for the sixth egg and you think: That egg has been in the refrigerator for some time and it could be rotten. What should you do?

Being a rational decision-maker, you may approach the problem by organizing the alternatives and states of nature as follows:

ALTERNATIVES

a_1—Break the sixth egg directly into the bowl containing the five eggs.

a_2—Break the sixth egg into an empty saucer, inspect it; if good, pour it into the bowl containing the five eggs; otherwise, throw it away.

a_3—Disregard the sixth egg and be satisfied with a five-egg omelet.

STATES OF NATURE

s_1—A good egg

s_2—A rotten egg

The probable outcomes Θ_{ij} are shown in Table 8-1 and are unevaluated.

A logical next step is to find an evaluation rule U to apply to each Θ_{ij} in order to compare resulting outcomes. Otherwise, it would be quite difficult to compare 0-, 5-, and 6-egg omelets and dirty saucers. This evaluation process is subjective and one must assume that an individual with proper background and

OTHER INVESTMENT CRITERIA

Table 8-1. A Decision Table for the Omelet Example

Alternatives	States of Nature	
	s_1 – good egg	s_2 – rotten egg
a_1 – Break sixth egg into bowl	6-egg omelet	0-egg omelet
a_2 – Break sixth egg into bowl	6-egg omelet and 1 dirty saucer	5-egg omelet and 1 dirty saucer
a_3 – Disregard sixth egg	5-egg omelet	5-egg omelet

experience is able to relate each outcome Θ_{ij} to some monetary value. For this example, let us assume the following monetary values.

Outcome	Monetary Value
Θ_{11} A six-egg omelet	$1.00
Θ_{21} A six-egg omelet and a dirty saucer	0.85
$\Theta_{31} = \Theta_{32}$ A five egg omelet	0.60
Θ_{22} A five-egg omelet and a dirty saucer	0.45
Θ_{12} no omelet	0.00

The evaluated decision table would now appear as shown in Table 8-2.

Table 8-2. An Evaluated Decision Table for the Omelet Example

Alternatives	States of Nature	
	s_1	s_2
a_1	$1.00	$0.00
a_2	.85	.45
a_3	.60	.60

You, as the decision-maker, now face the task of choosing a_1, a_2, or a_3. You may be very optimistic about the egg being good and ignore the possibility of a bad egg. In this certainty case of ignoring state s_2, you would undoubtedly select alternative a_1 because it has the greatest value. But in another case you may feel that both states of nature are possible.

Since we do not know the probability of each s_j in this uncertainty case, some arithmetic computation and mathematical thinking may help. In this case, one might let p represent the probability of s_1 and let $(1-p)$ represent the probability of s_2. Consequently, one may calculate expected values for each alternative as follows, where E symbolizes the expected value:

$$E(a_1) = 1.00p + 0.00(1-p) = 1.00p$$
$$E(a_2) = 0.85p + 0.45(1-p) = 0.40p + 45$$
$$E(a_3) = 0.60p + 0.60(1-p) = 0.60$$

With the exception of a_3, the expected value of the alternatives change as a function of p. Let us now develop a table of expected values by varying the probability p, assuming that p represents the probability that the egg is good, i.e., s_1 (see Table 8-3).

Table 8-3. Calculation of Expected Value of Each Alternative as a Function of the Probability of State One (s_1)

If p =	0.000	0.20	0.375	0.40	0.60	0.75	0.80	1.00
$E(a_1)$ =	0.00	0.20	0.375	0.40	0.60	0.75	0.80	1.00
$E(a_2)$ =	0.45	0.53	0.60	0.61	0.69	0.75	0.77	0.85
$E(a_3)$ =	0.60	0.60	0.60	0.60	0.60	0.60	0.60	0.60

The calculations show that the expected value of two alternatives are the same at three values of p: 0.375, 0.60, and 0.75. These three values are break-even points between the equivalent alternatives. Let us rearrange the data of Table 8-2 to identify optimum alternatives (see Table 8-4).

Each decision-maker approaches risk and uncertainty a little differently. But each decision-maker can reduce risk and uncertainty to a minimum by collecting some data and making a few calculations of the type shown in the omelet example. You may ask what types of information should be collected in an application environment. Suppose you are interested in buying an expensive numerically controlled machine tool. The alternatives might be an initial cost varying between $400,000 and $700,000. These amounts would represent that alternatives (a_i). The states of nature could be represented by quick obsolescence due to fast-changing technological improvements, an unstable market, a depressed economy, increased competition, and increased maintenance costs.

Table 8-4. Partitioning Variable p to Identify Optimum Alternatives

If p =	Then select Ai for i =	In order to obtain the maximum expected value
0 to 0.375	3	
0.375 to 0.75	2	
0.75 to 1.0	1	

An experienced decision-maker should be able to assess the alternatives in terms of dollars values and probabilities available to him and the states of nature he may encounter.

Additional Principles

A number of additional principles using probabilities are available to the decision-maker other than those previously discussed. These principles are:

1. The LaPlace principle in uncertainty decisions
2. The Wald principle in uncertainty decisions
3. The Hurwicz principle in uncertainty decisions
4. The Savage principle in uncertainty decisions
5. The replacement algorithms discussed in Chapter VII.

These principles will now be briefly discussed; example solutions will be presented later in the chapter.

THE LaPLACE PRINCIPLE. One of the oldest approaches to the uncertainty decision problem was suggested by the mathematician LaPlace, who was studying the mathematics of gambling. LaPlace reasoned that unless there was evidence to the contrary, one should assume that states of nature are equally likely and then evaluate the alternatives by their *expected values*. If this principle were applied to a machine purchase problem, then one might say that the life of a machine is somewhere between 10 and 20 years and that any number of years between 10 and 20 is just as likely as any other. The calculations for this particular principle might take the form of selecting four representative periods of machine life, *e.g.*, 5, 10, 15, and 20 years and then estimating the cost associated with each life.

Since the probability of each representative machine life is equal to that of the others, the expected value for each alternative is merely one-fourth of the sum of the cost associated with each machine life. Theoretically, the LaPlace principle states that since the probabilities associated with each state of nature are totally unknown for the m states, then the probability of each state should be assumed as equal to that of every other.

THE WALD PRINCIPLE. Another approach to this uncertain future is to find a choice which will assure one of doing as well as possible in spite of the darker aspects of the future. This approach is called the Wald principle. Basically, it states that one should select the alternative which has the best *minimum* value. Stated another way, it says that one should *maximize the least goodness* or *minimize the worst badness.* Consequently, the idea is also known as maximin or minimax, as appropriate.

One would apply this principle in practice by looking over the consequences of each future and then selecting the alternative in which the worst future is brightest. Operationally, this principle involves determining the maximum equivalent cost for each alternative and selecting the alternative with the least maximum equivalent cost.

THE HURWICZ PRINCIPLE. One might argue, as did Einstein, that nature may not be accommodating, but it certainly is not vindictive. Consequently, one ought not to consider only the dire consequences of an alternative. This thinking led Hurwicz to the conclusion that a reasonable evaluation of an alternative was *a value somewhere between the worst future and the best future,* depending upon the particular decision. Hurwicz argued that the decision-maker could identify his "optimism level" with regard to the particular decision by selecting a value for α, where α varies between $0 < \alpha < 1$. The value of α selected is the decision-maker's quantified judgment of the possible swing of nature toward the best consequences; the more optimistic the decision-maker, the higher the value of α he selects. Operationally, the principle simply states that the best future of each alternative is multiplied by α and the worst future is multiplied by $1 - \alpha$, and the sum of these two products provides a mixed future value. The principle further states that one ought to select that alternative which provides the best mixed future, which is the lowest value in the case where costs are used and the highest value in the case where benefits are used.

THE SAVAGE PRINCIPLE. One other method for uncertainty situations, orginated by L. J. Savage, involves the notion of regret. This idea might be stated, "If one picks an alternative and later finds that another alternative known at the time of the decision turns out to be better, then he will regret his choice."

This principle, therefore, works on the basis of making one's future regret as small as possible. To illustrate this, suppose one selected a machine that would last for 20 years, and found out later that it was replaced at the end of 10 years. In this case, one would regret the difference in cost between using the machine for 20 years and using it for 10 years. The regret or Savage principle goes on to state that one ought to select *that alternative in which the maximum regret is least*; consequently, the principle is often referred to as the minimax regret principle, *i.e.*, minimizing the greatest opportunity loss.

Examples of Other Evaluation Methods
In order to discuss the different problems that have uncertain or fallible information, let us look at a typical situation and consider it from the perspective of the various principles previously discussed, using the present worth method for evaluation.

OTHER INVESTMENT CRITERIA

THE PROBLEM. A company must decide between two machine designs, a_1 and a_2. The company's engineering staff estimates the first cost, annual disbursements, expected useful life, and estimated salvage value at the end of its estimated life for both alternatives as follows:

	a_1	a_2
First Cost	$50,000	$110,000
Annual Disbursements	$10,000	$5,000
Expected Life	10 years	20 years
Salvage Value	$26,000	$20,000

The company used straight-line depreciation and a 20% rate of return before taxes on its investments.

LaPlace Principle Application. This principle, in brief, states that unless there is evidence to the contrary, we must assume that all possibilities are equally likely. If this assumption is applied to the machine problem, it could be said that the life of the machine is somewhere between 10 and 20 years and any number between 10 and 20 is just as likely as any other. The basic assumption, then, is that a machine will be required for at least 10 years.

But there is no certainty that machine a_1 cannot function beyond 10 years of service life. Likewise, machine a_2 may not last 20 years. Suppose we calculate the present worth of costs for each alternative, assuming now that machine a_1 can function for 20 years as can machine a_2, using a spacing of 2 years between 10 and 20 years.

Machine a_1. Calculate the present worth of salvage values and disbursements. This is accomplished by using, for this example, straight-line depreciation and 20% present-worth factors made available from almost any engineering economics textbook. The calculations for machine a_1 are outlined in Table 8-5.

Note that salvage values are negative when costs are positive. Similar calculations for machine a_2 are summarized in Table 8-6. Since the LaPlace principle assumes all possibilities equally likely, then:

Machine a_1 Expenditures = 1/6 (87,721 + 92,011 + 94,800 + 96,672 + 96,764 + 98,648) = $94,619;

Machine a_2 Expenditures = 1/6 (120,426 + 125,912 + 129,300 + 131,594 + 132,970 + 133,838) = $129,011

Thus, the decision maker would probably choose a_1 since the present worth of expenditures is $34,392 less than a_2, *i.e.*, $129,011 − $94,619. Of course, one is asked to predict possible expenses for some time in the future, predictions that may not be too accurate. Nevertheless, long-range planning is a fact of industrial life and planners should be encouraged to use all available tools at their disposal.

ECONOMICS OF MACHINE TOOL PROCUREMENT

Table 8-5. Present-worth Calculations for Salvage Values and Disbursements (a_1)

Machine a_1	Years					
	10	12	14	16	18	20
1. Salvage values	$26,000	$21,200	$16,400	$11,600	$6,800	$2,000
2. Present worth of salvage value at year zero	−4,199	−3,379	−1,310	−628	−256	−52
3. Present worth of disbursements at year zero	41,920	44,390	46,110	47,300	48,120	48,700
4. First Cost	50,000	50,000	50,000	50,000	50,000	50,000
Total Present Worth (items 2, 3, and 4)	$87,721	$92,011	$94,800	$96,672	$97,864	$98,648

Table 8-6. Present Worth Caluclations for Salvage Values and Disbursements (a_2)

Machine a_2	Years					
	10	12	14	16	18	20
1. Salvage values	$65,000	$56,000	$47,000	$38,000	$29,000	$20,000
2. Present worth of salvage value at year zero	−10,498	−6,283	−3,755	−2,056	−1,090	−522
3. Present worth of disbursement at year zero	20,960	22,195	23,055	23,650	24,060	24,350
4. First Cost	110,000	110,000	110,000	110,000	110,000	110,000
Total Present Worth (items 2, 3, and 4)	$120,462	$125,912	$129,300	$131,594	$132,970	$133,828

OTHER INVESTMENT CRITERIA

Wald Principle Application. The Wald principle dictates that the decision-maker pick the maximum present cost that is least. Using the cost figures just developed, the maximum cost for machine a_1 is $98,648 and the corresponding cost for machine a_2 is $133,828. Therefore, the decision-maker would probably choose machine a_1 because it has the least maximum cost.

Hurwicz Principle Application. The Hurwicz principle advises choosing the best mix between the best of the bad aspects and the best of all worlds. Technically, this is done by selecting two futures from each alternative: (1) the best future and (2) the minimax future.

The minimax future is the same one selected in the Wald principle. After selection, these are mixed together to reflect one's own view of the future. In order to do this, a mixing constant must be selected. This mixing constant, sometimes called the coefficient of optimism, can be represented by α. Alpha can be as great as 1 and as small as 0. The larger the value of α, the greater the expressed optimism. Technically, this is accomplished as follows:

Alternative = α (best future) + $(1 - \alpha)$ minimax

Assume = 0.75

Then,

Machine a_1 = 0.75 ($87,721) + 0.25 ($98,648)
   = $65,791 + $24,662
   = $90,453;

Machine a_2 = 0.75 ($120,462) + 0.25 ($133,828)
   = $90,347 + $33,457
   = $123,804

The decision-maker would probably pick the alternative with the best mixed future, *i.e.*, machine a_1.

Savage Principle Application. Savage states that one's future regret should be made as small as possible. Technically, this is accomplished for our example by:

First, listing the various estimated lives of the machine and their present-worth amounts. For this purpose, we can use the tables developed for the LaPlace illustration.

Machine	Years					
	10	12	14	16	18	20
a_1	$ 87,721	$ 92,011	$ 94,800	$ 96,672	$ 97,864	$ 98,648
a_2	$120,462	$125,912	$129,300	$131,300	$132,970	$133,838

Second, by subtracting the smaller value in each column from each value in the column, we obtain:

Machine	\multicolumn{6}{c}{Years}
	10	12	14	16	18	20
a_1	0	0	0	0	0	0
a_2	$32,740	$33,901	$34,500	$34,922	$35,106	$35,180

We find that the largest regret for a_1 is zero and the largest for a_2 is $35,180. Therefore, the decision-maker's choice ought to be a_1 because if the worst happens it hurts the least.

The decision in actual applications would not be as obvious as in the example shown. In the interest of simplicity, disbursements and depreciation were considered constant over time. Actually, these categories of cost could vary widely throughout a future planning horizon.

Table 8-7 shows the various present worth savings between the two alternatives.

Table 8-7. **Summary of Present-worth Savings**

Principle	Selects	Present Worth of Savings
1. LaPlace	a_1	$34,392
2. Wald	a_1	35,180
3. Hurwicz	a_1	33,351
4. Savage	a_1	35,180

Retirement and replacement decisions for machinery and equipment are becoming increasingly important in the management field. The reader is referred to the replacement algorithms discussed in Chapter VII.

INDEX

Adverse minimum, p. 85
After tax cash inflows, p. 3
Algorithms, replacement, pp. 78, 83, 85
Annuity plan, p. 22
    discount calculations, p. 24

Business property taxes (*see* Taxes)

Capital budgeting, decisions of, p. 53
Capital budgeting methods, p. 55
Capital recovery factors, p. 84
Cash flow
    analysis of differences, p. 57
    discounted, pp. 2-3
    management of, p. 71
Compound interest tables, p. 4
Costs
    annual operating and maintenance, p. 1
    capital recovery, p. 1
Cutoff rate, p. 53

Data collection, pp. 4-5
Decision, formal representation of, p. 107
Declining balance,
    calculations, p. 33
    fixed percentage, p. 31
Depreciation
    double declining balanace, p. 31
    method of, p. 3
    methods, p. 29
    service output, p. 44
    straight-line, pp. 3, 29, 32
    sum of the years' digits, p. 35
    switch-over, p. 41
Deterioration, p. 85
Direct savings, p. 2
Discount, simple, p. 9
Discounted cash flow (*see* Cash flow)
Dynamic Equipment Policy, p. 85

Engineering Economy, p. 44
Equivalence, p. 14
Executive Decision and Operations Research, p. 53

Financial ratios, p. 71
Funds, cost of, p. 84

Gillette formula, p. 44

Hurwicz principle, p. 115

Indirect savings, p. 2
Interest
    compound, pp. 7, 11
    compound, conversion periods, p. 12
    nominal and effective, p. 13
    rate of return (trial and error), p. 67
    simple, pp. 7-8
Investment criteria, p. 107

LaPlace principle, p. 111

MAPI (*see* Algorithms)

Numerical control
    drilling machines, economics, p. 4
    justification, procedure, p. 1
    justification, cost graph, p. 2
    justification, trial and error, p. 4
    literature, p. 4
    problems in first purchase, p. 4
    versus conventional machine tools, p. 69

Obsolescence, p. 85
Operating inferiority, p. 85

Parameters of NC cost analysis, p. 4
Payout period, p. 65
Present worth, pp. 64, 66-67, 84
Profitability, p. 76
Programmed expenditures, p. 53
Project tax rate, p. 61

Real property taxes (*see* Taxes)
Repayment plans, p. 16
Replacement example, p. 86
Return
    calcuation of rate of, p. 68
    computing the rate of, p. 82
    on average cost, pp. 63, 65
    on first cost, pp. 63-65
    on investment, true rate, p. 64
Rate of return principles, p. 63
Ratio
    current assets to current liabilities, p. 76
    current liabilities to net worth, p. 77
    fixed assets to net worth, p. 76
    net profit to net worth, p. 76
    total debt to net worth, p. 76
Risk, p. 84

Savage principle, p. 115
Service output depreciation (*see* Depreciation)
Sinking fund, p. 22
Small Business Tax Act of 1958, p. 45
Straight-line depreciation (*see* Depreciation)
Study period, p. 59
Switch-over depreciation (*see* Depreciation)

Taxes
    business property, p. 62
    effect on capital budgeting, p. 60
    profit based, effect of depreciation method, p. 47
    real property, p. 62

Wald principle, p. 115